Scaffolded Minds

Integration and Disintegration

Philosophical Psychopathology

Jennifer Radden and Jeff Poland, editors

Scaffolded Minds: Integration and Disintegration, Somogy Varga (2019)

Real Hallucinations: Psychiatric Illness, Intentionality, and the Interpersonal World, Matthew Ratcliffe (2017)

Extraordinary Science: Responding to the Current Crisis in Psychiatric Research, Jeffrey Poland and Şerife Tekin, editors (2017)

Disturbed Consciousness: New Essays on Psychopathology and Theories of Consciousness, Rocco J. Gennaro, editor (2015)

Being Amoral: Psychopathy and Moral Incapacity, Thomas Schramme, editor (2014)

A Metaphysics of Psychopathology, Peter Zachar (2014)

Classifying Psychopathology: Mental Kinds and Natural Kinds, Harold Kincaid and Jacqueline Sullivan, editors (2014)

The Ethical Treatment of Depression: Autonomy through Psychotherapy, Paul Biegler (2011)

Addiction and Responsibility, Jeffrey Poland and George Graham, editors (2011)

Brain Fiction: Self-Deception and the Riddle of Confabulation, William Hirstein (2004)

Imagination and Its Pathologies, James Phillips and James Morley, editors (2003)

Scaffolded Minds

Integration and Disintegration

Somogy Varga

The MIT Press
Cambridge, Massachusetts
London, England

This book was set in ITC Stone Serif Std and ITC Stone Sans Std by Toppan Best-set Premedia Limited. Printed and bound in the United States of America.

Library of Congress Cataloging-in-Publication Data

Names: Varga, Somogy, 1974- author.
Title: Scaffolded minds : integration and disintegration / Somogy Varga.
Description: Cambridge, MA : The MIT Press, [2019] | Series: Philosophical Korpsychopathology | Includes bibliographical references and index.
Identifiers: LCCN 2018050125 | ISBN 9780262042628 (hardcover : alk. paper)
Subjects: LCSH: Philosophy of mind. | Cognitive science.
Classification: LCC BD418.3 .V37 2019 | DDC 128/.2--dc23 LC record available at https://lccn.loc.gov/2018050125

10 9 8 7 6 5 4 3 2 1

Contents

Preface and Acknowledgments

The impetus for this book arose from observing several intriguing changes in the contemporary philosophical landscape. We are witnessing a growing engagement in *empirically informed philosophy of mind*, which offers fertile interfaces between philosophy and cognitive science, allowing for the application of philosophical resources to the subject matter of scientific inquiries and their calibration in light of empirical findings. Moreover, a considerable amount of the philosophical work in this book was propelled by two distinct shifts within empirically informed philosophy of mind, which invite philosophical engagement with the theoretical commitments of emerging conceptual frameworks and research programs.

The first shift can be described in terms of an intensified study of the *embodied mind*. The theoretical basis of cognitive science no longer exclusively relies on cognitivist approaches that comprehend mental processes as abstract formal processes or as activation patterns in neural networks that can be adequately described in abstraction from the body and the environment. In contrast, embodied accounts subscribe to the idea that cognition often is best comprehended as the artifact of a dense interaction between neural and non-neural entities and processes. This is explanatorily relevant and exerts a profound influence on cognition in a way that stretches far beyond providing input into a cognitive system it remains closed off from.

The second shift can be described in terms of an intensified study of the *disordered mind*, expressing an acknowledgment of the convergence of the explanatory concerns of psychiatry and interdisciplinary inquiries in the mind. An empirically informed reflection about psychopathological phenomena constitutes a valuable resource for testing theories, offering real-life cases instead of the hypothetical ones that feature in philosophical thought experiments. Experimental manipulations in controlled environments can be fruitfully combined with studying "naturally" occurring changes in individuals with mental disorders in ecologically valid environments.

Combining these two shifts offers the possibility of complementing contributions and distinctive insights into cognitive processes, as well as the exploration of potential benefits for the understanding and treatment of mental disorders. A more complete understanding of our cognitive lives requires considering their dependence upon features of the (non-neural) body and environment and their vulnerability to malfunction. Moreover, if it is true that cognition is "embodied" in a nontrivial sense, then we may anticipate that increased attention to the body will yield epistemological gains in understanding how the mind works and harbor potential implications for the diagnosis and treatment of mental disturbances and disorders.

The overall aim of this book is to help create synergy at the intersection of embodiment and psychopathology. The book will motivate and defend the *actively scaffolded cognition* (ASC) framework, which restructures and repositions embodied approaches to promote a direct interdisciplinary dialogue among philosophy, psychiatry, and cognitive science. It will offer a taxonomy of ways in which cognition is scaffolded onto the body and the environment, and it will demonstrate that ASC can offer useful resources for comprehending prominent features of mental disorders and for providing new ideas for therapeutic measures.

I have benefited enormously from the writings of Tony Chemero, Andy Clark, Shaun Gallagher, Susan Hurley, Dan Hutto, Mark Johnson, Richard Menary, Mark Rowlands, John Sutton, Kim Sterelny, and Mike Wheeler, all of whom have established productive dialogues between the empirical sciences of the mind and philosophical inquiry. I am grateful to Philip Laughlin, senior editor at MIT, and series editors Jennifer Radden and Jeff Poland for support and encouragement from the start and for putting up with the delayed completion of this book. The constructive comments of four anonymous reviewers helped me to improve the manuscript. I have benefited greatly as well from continuous conversations with Remy Debes, Shaun Gallagher, David M. Gray, Thor Grünbaum, Detlef Heck, Dan Hutto, Beate Krickel, Albert Newen, Jennifer Radden, Heidi Samuelson, John Tienson, Deborah Tollefsen, and Michael

Wheeler, whose thoughtful comments and questions helped increase the lucidity and precision of some of the main arguments put forward in the book. Of course, these colleagues and friends bear no responsibility for the theses developed and defended in this book; any remaining errors are my own. Heidi Samuelson's outstanding copyediting considerably ameliorated the text, sparing the reader from chewing through passages overly inspired by various arrangements of non-English sentence structures.

The completion of the research for this project was made possible by a professional development assignment and a faculty research grant, both awarded by the University of Memphis. Furthermore, I've also benefited from support as a visiting research fellow at the Center for Mind, Brain and Cognitive Evolution at Ruhr Universität Bochum and as an affiliate faculty member of the Institute of Intelligent Systems at the University of Memphis.

Some of the material in the book draws on published journal articles. Thanks to the publishers for the permission to use material from the following pieces: "Embodied Concepts and Disordered Minds," *Journal of Medicine and Philosophy* 43 (2) (2018): 241–260; "Demarcating the Realm of Cognition," *Journal for General Philosophy of Science* 49 (3) (2017): 435–450; "Interaction and Extended Cognition," *Synthese* 193 (8) (2016): 2469–2496; "Cognition, Representations and Embodied Emotions: Investigating Cognitive Theory," *Erkenntnis* 79 (1) (2014): 165–190.

1 Philosophy, Psychiatry, and Cognitive Science

1.1 Empirically Informed Philosophy of Mind

Intensified interaction with the empirical sciences has led to numerous transformations in philosophical work. The traditional way of conceiving the task of philosophy is tied to a large extent to the method of conceptual analysis, which is applied to several central concepts in various fields of inquiry. For example, some philosophers of mind think that the chief task is to provide a fine-grained, purely a priori analysis of, say, folk-psychological concepts like belief and desire. In this view, philosophical analysis is not answerable to empirical facts, and the results of the inquiry are mainly attained based on mapping connections within the conceptual scheme that constitutes the medium of thinking. Mapping the relevant relations may include designing particular thought experiments, which shed light on the tacit meaning of concepts under investigation and unearth implicit principles that can help clarify conflicting judgments. But overall, this view depicts philosophy as essentially being in the business of descriptive conceptual analysis, aiming, as P. F. Strawson (1992, 7) puts it, to "produce a systematic account of the general conceptual structure of which our daily practice shows us to have a tacit and unconscious mastery."

In the contemporary philosophical landscape, such a view of the tasks and methods of philosophical inquiry is becoming much less common, and major scientific fields of inquiry are now complemented by subdivisions of philosophy that specialize in investigating a range of questions pertinent to the subject matter. The success of cognitive science has surely been a motivating factor for philosophers to account for new findings and to adjust their theories, topics, and approaches. Philosophers investigating the mind now often draw on findings in the sciences of the mind, reaching conclusions based on empirically informed reflection instead of a priori methods. Accompanying this reorientation in philosophical theory construction, it is now relatively customary to comprehend armchair "data" as defeasible and to deploy a "wide" reflective equilibrium methodology (e.g., Graham and Horgan 1994). This does not necessarily require a complete break with traditional methods of armchair analysis, but it definitively involves a decisive impulse toward "naturalization," reducing the "cognitive wiggle room" by recourse to empirical research (Weinberg 2017). Without this empirical input, especially in areas in the philosophy of mind, the view is that armchair approaches risk "losing contact with the very phenomena they seek to illuminate" (Kornblith 2017, 159).

These roughly delineated modifications in philosophical thinking have contributed to the emergence of a dynamically evolving specialized field, which encompasses a number of productive interfaces between philosophy and cognitive science. Of course, philosophy may not seem to be manifestly present in the everyday practice of cognitive science. Nonetheless, empirically informed philosophy of mind fits naturally into an interdisciplinary field with a multitude of methodological approaches that has, since its beginnings, regarded philosophy as one of its participating disciplines. In fact, some argue that due to the nature and subject matter of cognitive science, there is "no impassable gulf between those cognitive scientists who are philosophers and those who belong in the other disciplines, and there is no sharp line between the issues proper to the respective areas" (van Gelder 1998b, 134; Grush 2002).[1]

However, although there are no sharp divisions, philosophers have a particular role to play, and the nature of their work—for instance, evaluating the virtues of competing theories and determining their underlying commitments—is distinct, neither clearly conceptual nor empirical (Thomasson 2014). Sorting out the details of this type of philosophical work is beyond the aims of this book, but we may say that it engages empirical material, clarifies concepts, interprets tested and untested hypotheses, and forms new hypotheses, some of which can be tested. The approach lends itself to addressing themes that do not (yet) lend themselves to a transformation into scientifically tractable questions and that have not (yet) reached a level of maturation at which they could be confirmed or refuted (van Gelder 1998b).[2] In this sense, cognitive science is, as Daniel C. Dennett puts it, "a land of plenty for philosophers," because many of its questions "are still ill thought out, prematurely precipitated into forms that deserve critical reevaluation. If philosophy is, as my bumper sticker slogan has it, what you're doing until you figure out just what questions to ask, then there is a lot of philosophy to be done by cognitive scientists these days" (Dennett 2009, 232).

Moreover, this approach exhibits three important characteristics that are of central importance for the aims of this book. First, such empirically informed philosophy avoids the pitfalls of (cognitive) *scientism* (philosophers generate questions but should leave the answers to "proper" cognitive scientists) and *isolationism* (cognitive science has nothing to offer to philosophical analyses of conceptual schemes; for a discussion, see, e.g., Davies 2005; Ludwig 2015). Second, it is an *interactive* specialized subdivision, in the sense that its investigations create an interface that allows for combining the application of philosophical resources to the subject matter of scientific inquiries with the calibration of philosophical approaches in light of empirical findings and scientific accounts. Third, it is *naturalistic* in the sense that philosophical investigations are understood as continuous with empirical work in relevant fields. Unlike traditional approaches, empirically informed philosophy of mind holds that metaphysics should be informed by and be continuous with science.

1.2 Philosophy *in* Cognitive Science

Another way to help clarify the approach of this book is to explore how it corresponds (and fails to correspond) to extant distinctions in the literature. To distinguish empirically informed philosophy of mind from other philosophical work, some differentiate between philosophy *of* cognitive science and philosophy *in* cognitive science (e.g., Brook 2009). Empirically informed philosophy of mind is a way of doing philosophy *in* cognitive science—for instance, by offering integrative interpretations of tested hypotheses into larger frameworks, generating new hypotheses, and providing fine-grained conceptual clarifications. Philosophical work of this kind can be seen as philosophy *in* cognitive science because it constitutes a part of the practice of cognitive science, whether or not it is performed by researchers trained in philosophical methods. In contrast, philosophy *of* cognitive science can be pursued in several ways. For instance, it can be pursued by using the tools of philosophy of science, working out how general problems in the epistemology and metaphysics of scientific inquiry manifest themselves in cognitive science (Bechtel 2009, 2010; Bechtel and Herschbach 2010). Systematic considerations of the nature of explanation, confirmation, validity, the relation between theory and data, reduction, and so on in the special case of cognitive science are indispensable for securing scientific progress, especially in light of the different fundamental methodological and conceptual commitments of various disciplines comprising cognitive science (Samuels, Margolis, and Stich 2012).[3]

Although the type of empirically informed philosophy of mind that this book engages in can be understood largely as philosophy *in* cognitive science, it also involves significant amounts of reflection that may be more characteristic of philosophy *of* cognitive science. For example, investigating whether cognitive models fit behavioral data or whether certain correlations perhaps reflect causal or constitutive relations involves deliberation about the types of explanations that one takes to be suitable for cognitive processes. Moreover, offering a theoretical framework that aims to integrate various positions with emerging empirical findings involves reflection on the extent to which a single, unified view is possible, especially in light of the considerable complexity of mind and behavior.

1.3 Two Shifts

Having briefly delineated the nature of the empirically informed philosophy of mind that this book employs, it is important to note that the overall project is propelled by major theoretical reorientations in the field. This is the type of situation in which philosophical work is especially called upon to advance research by engaging often ambiguous theoretical commitments of emerging conceptual frameworks and research programs. This book brings together what can be seen as two shifts in the empirically informed study of the mind.

1.3.1 The Embodied Mind

The first shift is in the theoretical basis of cognitive science toward the *embodied mind*. Some maintain that we are witnessing a paradigm shift toward *embodied cognition* (EC), undermining the central idea of *cognitivism*, according to which "cognitive mental processes are operations defined on syntactically structured mental representations that are much like sentences" (Fodor 2000, 3–4). Of course, one may point out that somewhat similar shifts have occurred previously—such as during the 1980s, when emphasis progressively shifted from abstract formal descriptions of cognitive processes to connectionist approaches based on neural models of cognitive architecture and neural-based computation (Bermúdez 2014, 59–82).[4] However, though this shift has certainly introduced significant theoretical modifications, it left certain fundamental commitments untouched. One of these is a basic understanding of the mind that underlies cognitivism's methodological approach. Although the organism's body and sensorimotor systems deliver sensory input and enable behavioral output, the view is that they do not shape cognitive processing in any interesting and epistemologically significant way. This means that whether mental processes are best seen as abstract formal processes or as activation patterns in neural networks, the shared commitment is that they unfold inside brains and can be adequately explained in abstraction from the body and the environment.

In contrast, EC aims to provide an approach that accentuates the function of the body and the environment in cognition. EC accounts converge on some version of the general view that cognition is the product of the dynamic interaction of neural and non-neural processes. Without there being essential gaps between cognition, body, and environmental features, the idea is that numerous aspects of an agent's cognition profoundly depend upon features of that agent's non-neural body and environment. Before going further, it is worth pausing to specify the sense in which *dependence* is understood.

First, though logical dependence obtains between propositions, we are in this context concerned with *ontological dependence*, which describes a variety of relationships between entities or beings. Still, the sentence "cognition ontologically depends on the body" can be comprehended in several ways. Understood in a strict sense of ontological dependence, cognition could not exist if the body did not exist or did not supply its machinery with oxygen. Existential dependence in this strict sense is a relatively trivial relation that has little explanatory power in our context. Much more relevant for the purposes of this book is a different sense of dependence that we could call *nature dependence*. In this relation of dependence, cognition depends on the body not merely for its existence, but also for its nature or character. Saying that X (cognition) depends on Y (the body or environment) is in this sense roughly equivalent to saying that X is generally shaped by the specifics of Y or that aspects of X reflect aspects of Y. This relationship is important from an explanatory standpoint: when the nature or character of X is dependent on Y in this sense, then the characteristics of Y cannot be ignored in the explanation of X.

EC is best seen as a conceptual umbrella for several relatively un-unified research endeavors that have nevertheless influenced theory and practice in cognitive science. EC comprises labels like *embodied, embedded, extended,* and *enacted* cognition, which all endorse *epistemological inseparability* and thus the view that the organism's body and sensorimotor systems actively participate in the execution of cognition, such that we cannot provide a full understanding of cognitive processes by studying exclusively what is occurring inside the head of the cognizer.

1.3.2 The Disordered Mind

The second shift is toward more sustained study of the *disordered mind.* The vulnerability of our minds to division and self-alienation has long captivated philosophers, but the intense and puzzling perceptional, cognitive, emotional, and behavioral characteristics in mental disorders are now receiving a more prominent emphasis. Within the last three decades, the new interdisciplinary field of the philosophy of psychiatry and cognitive psychopathology has begun to develop and flourish (Fulford 2000; Fulford et al. 2003; Graham 2009).[5] Inspired by both Anglo-American analytic and continental philosophical traditions, philosophers, psychologists, and psychiatrists working in this area are striving to attain a more profound understanding of both psychiatric conditions and mental healthcare.

Within the philosophy of psychiatry, we may distinguish between overlapping areas of special interest (Graham and Stephens 1994; Murphy 2008), employing the distinction between philosophy *in* and *of* psychopathology. The latter includes fields of inquiry that deal with the way in which general problems in the philosophy of science related to explanation and classification manifest themselves in psychiatry as a special science. It also includes systematic reflections on how ethical considerations about rationality and self-determination apply to mental disorders. The former is progressively recognized to complement philosophy *in* cognitive science. For instance, philosophical reflections on questions surrounding personal identity have profited from analyses of mental disorders (K. V. Wilkes 1988; Humphrey and Dennett 1989).

Compatibly with this second shift, there is an increasing acknowledgment of the remarkable convergence of the concerns and approaches of cognitive science, psychiatry, and philosophical inquiries into the mind (see Cratsley and Samuels 2013). Psychiatrists have become increasingly attentive to the explanatory potential of cognitive science, perhaps assisted by the perceived lack of progress in the neurobiology and genetics of mental disorders (Bortolotti and Broome 2009; Kendler 2008; Kendler, Zachar, and Craver 2011), while cognitive scientists have gradually turned to exploring features of psychopathology, hoping to shed light on puzzling phenomena and to gain deeper insight into "normal" functioning. The view is that aspects of conditions classified as mental disorders can—not entirely unlike manipulations in experimental settings—provide opportunities to discern the nature of the mechanisms that underlie normal cognitive functioning. The study of the disordered mind thus supplements philosophy *in* cognitive science with empirically informed theorizing about psychopathological phenomena, which constitute valuable resources. Case studies from psychopathology offer important data that supports or challenges theories. In fact, the uniqueness of psychopathological aspects resembles actual instances of the sorts of imaginary conditions that appear in thought experiments, without raising concerns about plausibility.

1.4 A Productive Intersection

Although these two shifts in the empirically informed study of the mind may not be equally important and prevalent, they can complement each other in a number of ways.[6] Importantly, the intersection offers the possibility of distinctive insight. For instance, psychopathology offers a rich source of insight about the organizational, structural, and functional features of cognition and provides evidence relevant to the assessment of hypotheses. Experimental manipulations in controlled environments deployed to unearth these structures can be fruitfully combined with studying "naturally" occurring changes in individuals with mental disorders in ecologically valid environments. Models of cognition can be evaluated by exploring the extent to which they are able to explain cognitive performance in individuals with mental disorders and to offer viable explanations of the nature of the impairment. The study of psychopathology allows for observing arrangements of functional continuation and disruption in cognitive capacities, which enable inferences about the organization of normal functioning.

For instance, consider *empathy*, globally defined as the ability to understand and respond to others' mental states. Impairments in the capacity for empathy constitute a crucial dimension of several disorders. Studies on autism spectrum disorder (ASD), borderline personality disorder, and schizophrenia demonstrate that empathy is a mechanism that can be differentially impaired without, for instance, other damage in short- or long-term memory, showing that they function independently of each other. Once identified, this *single dissociation* can be used in epistemically constructive ways. For example, the existence of this single dissociation can lend initial support to the suggestion that empathy is not a unitary concept, but rather a multidimensional construct involving two distinct abilities: an emotional component and a cognitive component. In both ASD and borderline personality disorder, affective and cognitive empathy can be differentially impaired, such that a deficit in cognitive empathy is accompanied by a preserved emotional empathy (Smith 2009; Harari et al. 2010).

However, studies on alcoholism, controlled for psychiatric comorbidities, find the opposite pattern: impaired emotional empathy is accompanied by preserved cognitive empathy (Maurage et al. 2011). Taken together, the findings show a *double dissociation*, supporting the idea that emotional and cognitive empathy are two distinct abilities, likely reflecting two different underlying mechanisms. The now differentiated intact and impaired functions help generate a taxonomy of functional subsystems and offer information about the functional organization of the human mind, though without showing how these subsystems interact. The bottom line is that the experimental investigation of pathological dissociations opens new stimulating ways for the scientific investigation of cognition.

To the philosophical study of the mind and psychopathology, EC also offers some of the aspects that traditional cognitive science lacks with respect to crucial aspects of mental disorders. It has been argued that traditional cognitive science neglects the role of emotions, the body, and phenomenal consciousness. Take, for instance, three characteristic features of depression: motor retardation (e.g., slowed movements and speech, altered bodily awareness), emotional disturbances (e.g., low mood), and altered phenomenal quality (e.g., the sense of being captured in an unchangeable state), which are often portrayed in clinical and autobiographical descriptions (Ratcliffe 2008; Radden and Varga 2013). Of course, it is not the case that approaches that draw on the theoretical framework of traditional cognitive science deny that these disturbances are characteristic. The point is that they regard them as secondary to, and to a large extent caused by, cognitive biases in depression. This approach provides a potential target for cognitive intervention, but it comes at a relatively steep price. It is unable to account for the rich phenomenology of experience in depression and fails to suitably explain why pathologically altered states in depression are experienced as absolutely resistant to change and detached from the mental lives of others. This is where approaches that draw on EC could offer significant contributions to understanding mental disorders, analyzing them not merely as "brain dysfunctions" but as disturbances of an immersed embodied interaction with the environment, mediated by the brain.

1.5 Scaffolded Minds and Actively Scaffolded Cognition

The book can be described as combining two epistemological assumptions that arise from these two shifts. The first assumption is connected to the embodied nature of human minds, whereas the second assumption is connected to their fragile nature. This leads to the *epistemological conjecture* that though many psychological mechanisms are puzzling and unknown, we cannot provide a full understanding of cognition without studying (a) how cognition depends upon aspects of the non-neural body and environment and (b) how it is vulnerable to malfunction. The combination of these two assumptions leads to a fertile intersection with noteworthy epistemic potentials for philosophical research and clinical practice.

In light of the potential synergies and complementing contributions these two shifts offer, this book will investigate how they can be brought to work together and to explore potential benefits for the understanding and treatment of mental disorders. For this aim, the book will offer a theoretical framework, *actively scaffolded cognition* (ASC), which integrates a number of embodied approaches. The term *scaffolding* is beneficial for the aims of this book, as it offers a suitable amount of conceptual flexibility required for the task. The notion of cognitive scaffolding, originally advanced by Lev

Vygotsky and further developed by Kim Sterelny and others, became influential in developmental psychology, broadly designating support structures that enable a child to complete cognitive tasks that she could not accomplish on her own. But while scaffoldings in this tradition are seen often as temporary outside aids that are eventually removed once the child can perform a given task independently, the way the term is used in this book captures that they often become more or less permanent reinforcements of our cognitive machinery. In this way, the term will offer flexibility on several levels, which will help taxonomize forms of scaffolding and outline a view that is able to understand various embodied approaches as continuous with the idea of "cognitive niche construction" (e.g., Sterelny 2010).

This conceptual flexibility will assist in integrating accounts that subscribe to the principle of epistemological inseparability and that propel the study of mental disorders. ASC will offer a taxonomy of active scaffoldings, comprising two forms of *intrasomatic scaffolding* (simple and complex) and a specific form of *extrasomatic scaffolding*. The taxonomy is largely guided by pragmatic considerations linked to the explanation of common symptoms in mental disorders.

1.6 The Structure of the Book

Consistent with the principles of empirically informed philosophy of mind, this book aims to complete a dual task: mapping and application/calibration. Consequently, it naturally falls into two parts. The first part (mapping) makes a distinctive theoretical contribution, whereas the second part (application/calibration) shows how fine-grained philosophical distinctions can be applied to and calibrated by empirical research in psychopathology. Not only does this type of dual task naturally arise at the intersection of cognitive science, philosophy, and psychopathology, but contributions of this kind are crucial for an interdisciplinary field like cognitive science because they help address issues that traverse multiple areas of inquiry and link diverse approaches to understanding the mind.

The first task (chapters 2–5) is to engage the theoretical commitments of EC and offer a platform for further investigation. This is an important step given that EC is not a unified area of research, and the various research projects usually subsumed under the EC label lack homogeneity and established definitions of central concepts (see, e.g., Wilson 2002). One major goal is to draw the contours of ASC. For this, chapter 2 will explore two main theses that traditional cognitivism embraces but EC rejects. This chapter will provide an overview, which will gloss over details of particular positions to draw broad contours that only become noticeable at a particular level of abstraction. It will describe the cognitivist paradigm in terms of the commitments of three partially competing approaches (classicism, connectionism, and mixed architecture), which share a commitment to particular internal procedures that process information from the environment to perform actions. Although these approaches also exhibit differences when it comes to questions about representations and cognitive architecture, they share a commitment to the two principles of the separability thesis, which define the "disembodied" picture that EC opposes. In addition, the chapter will discuss clinical cognitivism, which applies principles of cognitivism to a clinical and therapeutic context. Cognitivism is a relatively unified interdisciplinary approach that aims to comprehend the causal processes that execute computational operations on representational structures, but EC lacks the characteristics of a well-defined and unified theoretical approach. At least at this stage of its development, EC offers valuable corrections to the cognitivist approach to cognitive science, but it should probably not (yet) be seen as offering a full alternative to cognitivism.

With this background in place, chapter 3 will disentangle differing positions in current debates and introduce a number of fine-grained distinctions. The chapter will illustrate how the notion of ASC can be used to construct a framework that integrates a medley of positions that support at least one of the two claims of inseparability. The chapter will steer clear of some of the debates about the extent to which mind and world are intertwined and focus instead on offering a preliminary idea of what it means for cognition to be *scaffolded* onto the environment and sensorimotor processes. To exclude trivial forms of dependence, the notions of nature-dependence and active scaffolding will be introduced; additional details of ASC and its commitments will be further specified in chapters 4 and 5.

The flexibility of the concept of scaffolding at this stage will help to tailor a taxonomy of forms of scaffolding, which is guided by pragmatic considerations regarding common symptoms in mental disorders and will therefore neither offer a comprehensive ontology nor focus on a particular cognitive domain or function. With the aim to provide distinctions that could be productively applied to the context of psychopathology, the chapter will distinguish between two forms of intrasomatic scaffolding (simple and complex) and a specific form of extrasomatic scaffolding (intersomatic), which is a distinct kind of socially scaffolded cognition. Intra- and extrasomatic scaffolding thus are the classes of the ASC genus.

The taxonomy raises several weighty questions about the relationship of dependence that holds between the scaffold and what is scaffolded. Chapters 4 and 5 will confront questions concerning inseparability in the ontological sense, while also providing further specifications of other aspects of ASC. They will explore two different paths. Chapter 4 will draw on recent debates on the boundaries of cognition, while chapter 5 will address the issue in light of more general questions on how to delineate the boundaries of mainly biological systems and mechanisms. To further specify ASC and its underlying "global" commitments, chapter 4 will position ASC in broader discussions about the boundaries of cognition in the philosophy of cognitive science, which is intertwined with the question of ontological inseparability. This is a debate between cognitivism and EC, as well as between EC accounts, and numerous philosophers maintain that the discussion would greatly benefit from providing a "mark of the cognitive" that picks out all and exclusively cognitive processes. The chapter will first engage two accounts that distinguish cognitive processing from mere information processing. This is followed by an investigation of the more general, fine-grained versus coarse-grained debate about the question of the correct grain level that should be deployed to examine cognitive processes. Roughly put, the debate revolves around this question: Should the boundary of the cognitive be fixed by examining the fine-grained functional details of cognition, or should the focus be on larger cognitive ensembles?

Although it has been noted that the debate has reached an impasse characterized by a clashing of intuitions, chapter 4 will argue that the failure of trained philosophers to elicit converging intuitions about a circumscribed subject might indicate a deeper problem. The chapter will explore underlying sources of the opposing intuitions, argue that "cognition" may be a special kind of prototype concept, and yet oppose eliminating the concept from scientific taxonomy in favor of several less inclusive labels. Rather, there is perhaps an opportunity here to explore the possibility of a kind of pluralism that is familiar from mature sciences and that neither implies antirealism nor impedes scientific progress. The attempt is in part motivated by a suspicion that no single account of cognition is both broad and specific enough to account for the wide variety of cognitive processes. The broad range of interdisciplinary goals pursued in cognitive scientific inquiry and the complex nature of cognition and mind should render us suspicious of the idea that a single, unified framework will eventually explain the entire range of cognitive processes. Importantly, it is possible to embrace pluralism while still holding onto the effort to integrate relatively dissimilar positions into loosely knit frameworks. Clarity on such matters may help avoid confusion about whether novel approaches complement traditional cognitive science or provide an alternative to it.

Unlike the approaches discussed in chapter 4, chapter 5 will engage influential accounts of causal and constitutive relevance, aspiring to secure a more neutral base. The chapter will draw on an interventionist account of causal relevance (M) as defended by James Woodward (2003, 2010, 2015). It will add to M further conditions of stability and specificity to elucidate the particular relationship of nature-dependence and active scaffolding holding between the relata in ASC. The addition of these auxiliary conditions will lead to the manipulability account of active scaffolding, or *Active Scaffolding (M)*. M, thus enhanced, will capture many active scaffolding relationships; others are often characterized by mutual and bidirectional difference-making. This compels considering a metaphysically distinct constitutive relationship, as investigated by the mutual manipulability account (MM) of constitutive relevance (Craver 2007). Recent work on mechanistic explanation (Bechtel 2017; Bechtel 2008; Craver 2007; Machamer, Darden, and Craver 2000) will help rethink relations of dependence that ASC aims to capture and provide a mutual manipulability account of active scaffolding, or *Active Scaffolding (MM)*.

On a first pass, the chapter will use MM to explicate constitution as a difference-making relation, and it will deploy two of its conditions as a test for ASC. MM offers three advantages. First, it provides an independently motivated basis for specifying ASC, which avoids certain risks that chapter 4 will identify. Second, it dovetails with the general pragmatic ambitions of this book, which seeks to explore the possibilities for extending the range of therapeutic interventions. Third, it retains proximity to explanatory practices in the relevant scientific fields and offers considerations on amenability to experimental testing and confirmation. To avert the risk of importing some difficulties linked to the notion of intervention, the slightly revised MM* will be introduced, which relaxes some requirements on interventions.

The second task (chapters 6–8) is to show that ASC is a productive framework for considerations about a number of characteristic features in mental disorders. The focus will be on problems with altered bodily experience and social cognition deficits, which are characteristic of a wide range of mental disorders, and the task will be to apply and adjust various conceptual and theoretical resources of ASC. Embarking on such a project seems attractive in light of the numerous potential benefits for diagnosis and treatment. First, understanding how scaffolded processes can *disintegrate* at many different junctures and at many different developmental stages, with each combination leading to markedly different downstream consequences, may assist in comprehending heterogeneous behavioral symptomatology. Second, in light of the relatively modest efficacy of current treatment options, it is reasonable to explore scaffolding structures that perhaps eventually can be exploited for therapeutic purposes, complementing pharmaceutical and psychological interventions.

Embarking on such an enterprise, chapter 6 will offer an application of the idea that certain concepts and cognitive activities are, in various ways, scaffolded onto the sensorimotor system. More precisely, it will explore intrasomatic scaffolding and distinguish two ways to comprehend the scaffolding relationship that are consistent with the epistemological inseparability thesis. The chapter offers support for the ASC framework and analyzes several studies to provide a more comprehensive understanding of mechanisms that are involved in the symptomatology of depression—in particular, motor retardation. This work is accompanied by reflections on potential therapeutic implications and considerations about the kind of further research that would be needed for a systematic clinical application.

Chapter 7 will expand the scope of inquiry and address a particular form of intersomatic scaffolding, which not only involves the agent's sensorimotor apparatus but also extends into the environment to include external structures that adaptively guide behavior. In such cases, the vehicles of cognition appear to individuate externally, constituting a neural, bodily, and extrabodily assembly. The chapter will trace the development of interactive skills that enable behavioral synchrony, which sometimes drives cognitive processes in children and adults. The focus will be on dysfunctional emotion regulation in depression. Emotion regulation has a genuine cognitive function and serves as the basis of an "extended" regulation of emotional and physiological arousal. Improving our comprehension of this issue has important implications for theory and practice, and interactional synchrony might be a promising operational construct for studying diagnostic and therapeutic opportunities.

Chapter 8 will focus on pathological alterations in the skills that enable us to smoothly understand each other in various contexts. It will be shown that mindreading skills are supported by an epistemically engineered environment and scaffolded onto the human body and sensorimotor apparatus. But in that case, one might anticipate new avenues for understanding common problems with social cognition in mental disorders. Research in autism spectrum disorder (ASD), for instance, has largely disregarded the role of the body and movement and described social cognition impairments exclusively in terms of mindreading deficits. The chapter will aim to incorporate recent empirical work in light of the conceptual distinctions established in the first part of the book and will show how an analysis in terms of scaffoldings offers new perspectives on central features in ASD. Consistent with the epistemological inseparability thesis, the chapter will support the view that an adequate explanation of social cognition in ASD needs to consider possible sensorimotor impairments and their effect on disintegration in the higher-order functions that they assist.

Overall, the second part of the book aims to establish that shifting attention from mental symptoms to fine-grained sensorimotor aspects and further improving the position proposed in the book can lead to identifying diagnostic subtypes, or even to specific sensorimotor markers for early diagnosis. One great advantage of identifying such markers is that they lend themselves to noninvasive, objective measurement that is relatively independent of cognitive-linguistic abilities. After summing up what we have covered in the book, the conclusion will end by noting potential contributions to recent discussions on reduction in psychiatry.

In several ways, this book attempts to walk a thin line. It primarily appeals to philosophers, especially those focusing on empirically informed areas of philosophy of mind and (meta)theoretical issues, but also to mental health professionals interested in current reflection on the theoretical basis of the scientific study of the mind. The relatively pluralistic position might help readers from the latter group explore the discussion without requiring substantial prior commitments, but one could also worry that it results in a less sharp theoretical position than other contributors have adopted. However, the reasons for adopting this position are mainly theoretical. Although the ASC framework holds that explanatory purposes require adopting a taxonomy of various forms of active scaffolding, the choice of not endorsing a single, monolithic theoretical framework is supported by considerations about metaphysical disputes about the boundaries of cognition, as well as debates on pluralism in other fields. To be clear, ASC is not neutral—for instance, it denies cognitivist assumptions about separability—and it is not pluralist in the sense that it grants equal status to the entire range of theoretical approaches. Instead, the overall outlook on cognition is pluralist in the sense of granting cognitivism a place in the larger story about cognition while remaining skeptical of the view that there is a single theoretical framework that can adequately deal with the entire range of cognitive processes.

The book also walks a thin line when it comes to the relationship between the ASC framework and the empirical material discussed in the second part of the book. As a note of caution, it should be stressed that there sometimes will be a disconnect between the relative precision of the definition of manipulability in the framework of ASC and the empirical material explored in this book. Because the empirical material on this subject is relatively sparse, at least compared to other areas of psychology, clinical psychology, and psychiatry, some compromises are difficult to avoid. In some cases, perhaps somewhat imprudently, active scaffolding will be inferred from complementary but separate studies. Moreover, some of the findings mentioned, especially in the course of the last three chapters, are relevant for the overall goals of the book but have not been probed by a sufficient number of bottom-up and top-down interventions.

2 Cognitivism, Clinical Cognitivism, and Embodied Cognition

2.1 Information Processing

A little over a century ago, psychology was undergoing a transformation that other scientific disciplines, like physics and biology, had already completed. In a multifaceted process, it separated itself from philosophy and became an experimental science, without turning into a physiological subdiscipline. The architect of modern experimental psychology, Wilhelm Wundt, introduced the experimental method of introspection and created controlled circumstances to investigate the domain of conscious mental processes. Although experiments principally concerned sensations, Wundt's student E. B. Titchener increased the scope of the experimental method of introspection to include higher cognitive processes, which in his view involved content-bearing mental images. Trained introspection was widely regarded as the characteristic method of this new scientific endeavor and as crucial for qualitative analysis, but it also became increasingly accepted that systematic quantitative evaluation of behavior needs to complement introspection. Some have argued that to acquire and maintain a status as a scientific discipline, psychology must employ a methodology that strikes a balance between studying first-person experience and objectively observable features.

The plea for such a balanced methodology only encountered limited support. Instead, the emphasis on introspection as an experimental method decreased over the subsequent decades, and psychologists and psychiatrists began favoring a broadly behaviorist approach. Redefining psychology's methodology and subject matter, J. B. Watson and B. F. Skinner reframed psychology as the science of behavior that observes stimulus-response patterns. Psychology was no longer the science of the conscious mind, and in the 1950s—with the demise of behaviorism and the onset of the "cognitive revolution"—mental phenomena for the first time were rehabilitated as suitable subject matter for scientific psychology (Davies 2005). The cognitive revolution comprehended cognition as a more limited category than adaptive, context-sensitive behavior, stressing that cognition requires an extra component, standardly understood in terms of *information processing*. The emerging concept of *information*—referring to states or vehicles that carry representational contents—enabled researchers to focus attention on cognitive processing of information without having to fear the label of *mentalism*.

Over the following two decades, cognitivism established itself as a successor to behaviorism in a process that has been characterized as a Kuhnian revolution, offering a new paradigm for the research of psychological phenomena (Baars 1986). The revolution gave rise to an interdisciplinary field of scientific investigation of the mind that comprised research in linguistics, computer science, anthropology, psychology, neuroscience, and philosophy. Although the term *cognitive science* was not coined until the field developed some institutional anchoring in the 1970s (Norman and Rumelhart 1975), the birth of cognitive science is often associated with interdisciplinary meetings on emerging issues in theoretical linguistics, artificial intelligence, and information theory in the mid-1950s (Bechtel and Herschbach 2010). The institutional presence further developed the engagement in "an attempt to bind our joint enterprise even more tightly by a common umbrella name, Cognitive Science" (Newell 1980, 136). Nonetheless, from the very beginning, cognitive science was more than merely a common umbrella name and subscribed to a certain theoretical orientation, often referred to as *cognitivism*.

First, from its beginnings it focused on a specific part of mental lives, perhaps in part motivated by practical challenges. As Boden (2006, 11) notes, "Because motivation, emotion, and social interaction ... are even more difficult to study—and to simulate—than cognition is, they were soon put onto the back burner. They were left there for thirty years, while cognition got almost all the attention. Vast amounts of research were done on perception, language, problem-solving, concepts, belief, memory, and learning."

Second, cognitivism seeks to advance our understanding of *cognition* (understood in a broad sense as encompassing mental processing, including abstract reasoning, planning, memory, perception, language, and motor control) and to explain behavior, it aims to achieve this by providing an ever-better comprehension of representations and the causal processes that execute computational operations on them. This is clearly conveyed in the opening sentence of the preface to an influential volume: "Cognitive science is the study of mental representations and computations and of the physical systems that support those processes" (Bly and Rumelhart 1999, xv). Broadly, *computation* refers to drawing inferences based on representations (see Boden 2006). Although there are differences between computation and information processing that might matter in particular contexts (see, e.g., Piccinini and Scarantino 2010), they are to a large extent used interchangeably in the literature. Importantly, the view of cognition as essentially information processing became deeply entrenched in the research.

2.2 Cognitivism

The emerging enterprise faced important questions about the extent to which it is continuous with other natural sciences, as well as about the extent to which a cognitive system ought to be comprehended as continuous with other physical systems. The emphasis on information processing provided some answers to these questions. Crudely put, the view is that biological systems manipulate biological material to maintain homeostasis, and cognitive systems manipulate or process information to generate and regulate behavior.

But even so, conceptualizing the mind as essentially an information processor akin to a computer is not straightforwardly evident and requires justification. A full explanation of how researchers came to think of the activity of the mind in this manner would require detailed historical analysis, but there are a couple of contributing factors worth mentioning. First, the conceptualization of the mind as a computer generated explanatory insights into the maneuvers of cognitive processes and facilitated predictions of performance under specific circumstances. Although this growth in terms of explanatory and predictive power has at least in part served as a justification, the acceptance of the view was facilitated by the unparalleled rise and success of the digital computer and an influential tradition in philosophical thinking that characterizes the human species in terms of its ability to develop and use both linguistic (natural languages) and nonlinguistic (e.g., pictorial) representational systems.

Second, conceiving of cognition in terms of computation was also seen as assisting the naturalization of the mind and the demystification of vexing metaphysical questions. Thus conceived, cognitive activity has a "dual profile," being physical in some sense, as well as sensitive to reasons and beliefs (Clark 1997). Just like the computer, our brains could contain states that represent, for instance, the environment as being "cold," followed by state transitions realized as physical brain events that lead to a new set of states, the proper interpretation of which is "turn off the A/C." It is not difficult to see the appeal. If the mind is essentially software that runs on the hardware of the brain, then one can grasp how it can exist independently of its particular physical form without having to posit the existence of a nonphysical substance. Metaphysical questions about the existence of mental states become no more puzzling than the existence of relevant states in a computer. Cognitive processes in this view are no longer mysterious, but part of the physical order.

The main idea of cognition as computation can be worked out in several ways, depending on computer architecture (Tienson 1988). As Searle (1992, 198) notes, though cognitivists agree that "the brain is a computer and mental processes are computational," it is under debate "whether or not the brain is a digital computer of the old-fashioned von Neumann variety or whether it is a connectionist machine."[1] Along these lines, this chapter will continue by distinguishing "classical" and "connectionist" accounts. Only the former is committed to the idea of a "language of thought" and to representations with combinatorial features, but both accounts posit causally active, representational states and computational procedures and can be comprehended as research programs *within the same cognitivist paradigm.* Ahead, the focus will be on analyzing the main claims of these accounts, with particular emphasis on theoretical commitments that help elucidate the emergence of the embodied cognition (EC) research program as a reaction to cognitivism. This approach makes it easier to explore the rather inhomogeneous EC research program still in its nascence and to convey how EC, in various ways, rejects the view that cognition is separable from the body and one's environment. To provide a fuller picture, the last part of the chapter will present some of the main philosophical psychological roots and sources of inspiration that were instrumental in the development of EC.

2.2.1 Classical Cognitivism

The *representational theory of mind* understands intentional states as relations to mental representations and comprehends their intentionality in terms of semantic properties. Very roughly, it can be characterized along two dimensions. First, it portrays mental states in folk-psychological terms (perceptions, beliefs, desires, etc.), as intentional states that can be individuated by their content and that are evaluable along specific criteria. For instance, perceptions can be accurate or inaccurate, thoughts can be true or false, and desires can be appropriate or inappropriate. "Believing that x" is being suitably related to a mental representation with the propositional content "that x" (evaluable as true or false), whereas "desiring that x" is being suitably related to x in a different manner (evaluable as appropriate or inappropriate). Second, it postulates the existence of mental representations, which act as intermediaries between the cognizing organism and the world.[2]

Classical cognitivism develops some of these ideas and maintains that mental representations are symbolic and formal (Fodor 1981). They are not consciously accessible, but causally implicated in the storage and manipulation of informational structures that contribute to the formation of propositional attitudes (Thagard 2005). Information processing involves symbols, and thus entities that have *syntax* (physical properties like shape and form) but lack semantics, which only arises when we syntactically combine symbols. The idea of cognition as symbol-involving information processing is thus paired with a fundamental idea that computation operates with representational mental states that stand in for some external state of affairs. As Fodor's (1975) motto maintains, "no computation without representation."

Main tenets of the classical view can be illuminated by drawing attention to (a) the relation it posits between syntax and semantics, (b) its level of description, (c) the way it models the mind as a computer, and (d) the way it models representations on linguistic representations. Let us briefly review these aspects:

(a) Fodor (1987) maintains that syntactic properties are physical properties that are correlated with semantic properties. If that were not the case, so the reasoning goes, then their recombination would not allow *semantic productivity*, the formation of additional sentences with related, but distinctive, semantic properties. Thus the acknowledgment of the possibility of semantic productivity and compositionality implicates that representations exhibit syntactic properties. As Fodor (1987, 19) puts it, "Only symbols have syntax, and our best available theory of mental processes ... needs the picture of the mind as syntax driven machine."

(b) Cognition as processing information by manipulating symbols (representations) designates a level of description that is autonomous in the sense of being independent of neuroscientific knowledge about brain structures and processes (Fodor 1975; Newell 1990). The algorithms operate on a functional level, without reference to physical machine properties (Pylyshyn 1984). Scientific studies exploring neural organization and structure require an additional level of comprehension of information-processing tasks, which is independent of the mechanisms and structural features that implement them (see, e.g., Marr 1982). The fact that physically different but functionally identical mental representations can be multiply realized (in biological or nonbiological entities or systems) supports the view that neuroscientific evidence is unable to offer significant contributions to cognitive psychological investigations about the (functional) nature of mental representations.

(c) Classicists put the computer metaphor to work to model the correspondence between symbol structure and physical structure.[3] As an example, Fodor and Pylyshyn (1988, 13) argue that the relation "part of" that holds between two symbols presumably corresponds to a physical relation between neural states. Others maintain that computational systems in computers and brains should be comprehended as "physical symbol systems" and stress that the way physical symbol systems are conceptualized in artificial intelligence and computer science should be taken seriously in cognitive science (Newell 1980; Newell and Simon 1972; Newell 1976). They propose that a physical system can only be capable of intelligent action if it is a physical *symbol* system. Such a system is both material (realizable in the physical world) and capable of manipulating symbols. For instance, symbolic structures stored in the computer are transformed via purely formal rules. These rules are *syntactic* in the sense that they are connected only to the form of symbols, not their meaning. They are meaningless states belonging to a material system in which their function depends on their material properties. The semantic characteristics of the symbol designated for an object and the characteristics of that object stand in an arbitrary relation. A computer can solve a simple multiplication problem by transforming the input symbol "4 × 4" and producing the symbol "16" as output, using a formal rule that links one type of input to another type of output. It applies the syntactic rules while "understanding" neither the computed function nor the meaning of the symbols. The function of symbols is to represent and be combined with others into a physical symbol system based on rules, which can give rise to new symbols.

Such physical symbol systems are purportedly continuous with our minds: when the computer obeys such formal rules and manipulates symbols, it does so in a manner that would count as cognition if executed by a human being. The symbols it deploys "are in fact the same symbols that we humans have and use every day of our lives. Stated another way, the hypothesis is that humans are instances of physical symbol systems, and, by virtue of this, mind enters into the physical universe" (Newell 1980, 136). This hypothesis of continuity defined the interdisciplinary quest for a scientific theory of mind and offered an initial model for explaining cognition in terms of processes that operate on representations. As Newell (1980, 136) has put it, "The physical symbol system is to our enterprise what the theory of evolution is to all biology, the cell doctrine to cellular biology, the notion of germs to the scientific concept of disease, the notion of tectonic plates to structural geology." An important addition is, of course, that the brain is the physical seat of a representational system: a specific neural architecture is the "hardware" on which the "software" of the mind runs. This is in part what allowed cognitive psychology to remain neutral about neuroscientific research on biological details of the realization of cognition, focusing instead on studying cognition in terms of the types of algorithms digital computers could implement.

(d) The classical view models mental representations on linguistic representations, in part to explain certain characteristic properties of cognition, like being productive, systematic, and truth-preserving (Fodor 2000). Fodor (1975) influentially applied this model to understand cognitive processing in human beings and argued that cognition occurred in a "language of thought" in which representational vehicles are sentences constructed from symbols in accordance with a combinatorial syntax. The idea was thought to offer suitable explanation of a variety of prominent features in our psychological makeup: the ability to comprehend new thoughts and sentences (productivity), the fact that understanding a thought involves understanding connected thoughts (systematicity), and the ability to understand the meaning of sentences from the meanings of their lexical parts (compositionality). Fodor has not entirely settled the extent to which such an internal representational system should count as a language, in part because languages are conventional, whereas the language of thought is not. But he stresses that it works very analogously to representations in natural language. As he notes, "Representation presupposes a medium of representation, and there is no symbolization without symbols. In particular, there is no internal representation without an internal language" (Fodor 1975, 55). The system is physically realized in our brains, but it is not an overly intellectual medium unique to human beings: the representational systems that some animals employ resemble ours (Fodor 1975, 57), which is why the language of thought cannot be a natural language. Instead, natural language must presuppose a language of thought.

Overall, in the classical view, cognitive processes store and manipulate information-carrying representations, which are comprehended as mental objects that display semantic properties like reference and truth conditions. Cognition consists of sequences of tokenings of mental representations, in which (i) mental representations are the "objects" of propositional attitudes, (ii) operations on them are causally linked to their syntactic structure, and (iii) they are functionally describable entities that can be multiply realized by different physical properties of cognizers.

Whether the classical view correctly models how the mind works has been at the center of several complex debates, the depth of which we cannot do justice to here. Among the classical view's appealing features is that it provides a framework for programming computers that exhibit problem-solving capacities. Moreover, it is consistent with important efforts that aim to offer a naturalist account of the mind as a nonmysterious part of the physical world and explain how changes in a physical system are able to preserve semantic properties.

2.2.2 Connectionism

The arrival of connectionism in the 1980s has been variously portrayed as a "new wave" in cognitive science, a "paradigm shift" (Schneider 1987; Horgan and Tienson 1991), and a decisive breakthrough in understanding the mind. Connectionism has roots in the work of McCulloch and Pitts (1943) on interconnected logic gates as idealized neuronal models. Inspired by the growing knowledge of how the brain works, connectionism challenged some of the fundamental assumptions of the classical view (Rumelhart, McClelland, and Group 1986).

Although connectionism seeks an alternative to the classical conception, the ensuing talk about a Kuhnian "paradigm shift" appears overstated, in part because it is unclear to what extent the new paradigm can handle the explanatory successes of the old one (for a discussion, see McLaughlin 1993). Both classicism and connectionism model cognition with computer programs, but whereas the former relies on modeling that transforms symbols in a rule-based manner, the latter mimics the brain's learning processes by creating networks of artificial neurons that consist of nodes (input, output, and hidden) linked in a configuration of connections (Hinton 1992). In contrast to classical architecture, connectionist networks lack a *central executive*—a central processing unit running a stored program that regulates the system. Instead, nodes only receive information from connected nodes. Of course, activation patterns in several nodes can be seen as having representational content and as being able to preserve "knowledge" in the form of particular patterns of connections between nodes. Here's how Fodor and Pylyshyn, proponents of classicism, present some of the main differences:

> Classical models of the mind were derived from the structure of Turing and Von Neumann machines. They are not, of course, committed to the details of these machines as exemplified in Turing's original formulation or in typical commercial computers; only to the basic idea that the kind of computing that is relevant to understanding cognition involves operations on symbols (see Newell, 1980,

> 1982; Fodor 1976, 1987; or Pylyshyn, 1980, 1984). In contrast, Connectionists propose to design systems that can exhibit intelligent behavior without storing, retrieving, or otherwise operating on structured symbolic expressions. The style of processing carried out in such models is thus strikingly unlike what goes on when conventional machines are computing some function. (Fodor and Pylyshyn 1988, 5)

To this description, we may add that for classicism, computational processes are much like arithmetic, consisting of the transformation of symbols (1, 2, 3, etc.) according to rules for performing operations specified by other symbols (+, -, ×, etc.). For connectionism, computational processes require no stored symbols or operational rules, but consist of causal processes by which units excite and inhibit each other (Bechtel and Abrahamsen 2002, 2). Also, though classical computational theories remain neutral about neuroscientific research on biological details of the realization of cognition in the brain, connectionist models of cognition consider neurophysiological and neuroanatomical findings (McLeod, Plunkett, and Rolls 1998), and neurocomputational models are even more constrained by these findings (Churchland and Sejnowski 1992; Rogers and McClelland 2014). Connectionist functional architectures are inspired by preliminary comprehensions of how actual neural networks operate, opposing the view that the basic units of computation are symbols (Churchland 1995).[4] Unlike in the case of classical architecture, a connectionist system does without a central processing unit and exhibits only local connections between nodes, allowing for simultaneous local processing distributed over the entire system. Moreover, connectionists operate with the notion of *distributed representation*, conceived not as akin to words in a "language of thought" but as an activation pattern that unfolds across a cluster of units (see, e.g., Horgan and Tienson 1992b).

Modeling how synapses link neurons, the nodes in the connectionist system continuously pass valued activations between them and bear a weighted connection to other nodes that change in accord with a learning algorithm. Input nodes receive a particular level of activation, which then spreads throughout the network depending on the already existing activation pattern and the characteristics of the connecting links. Equations determine activation spread in time, and the output of the system is the output node network settling into a stable state. Changes in connection strengths lead to alterations in function, which is in part what allows networks to be trained to compute various input/output functions. Basically, such algorithms for supervised and unsupervised learning calculate the right weights for a particular function (Hinton 1992). Training a network to perform a task aims to coordinate the activities in the input units and the network such that the desired output is achieved. Upon determining how accurately the output of the network corresponds with the desired output, the next task is to modify the weights until a suitable approximation of the desired output is achieved. To take an example of how this can work, consider a sentence-prediction model with a basic recurrent network that can be trained on a succession of sentences (Elman 1991, 1993). The sentences used during the training include grammatical dependencies (e.g., subject-verb agreement) and a little more than twenty words. A word in the sentence functions as the input, and the target output of the model is the next word in the sentence. During the training, the weights are adjusted by a back-propagation algorithm. In the end, without any explicit grammatical rules, the model can predict plausible continuations for sequences (e.g., "cats chase dogs"), demonstrating that connectionist models can display productivity (Churchland 1995).

One reason for the popularity of connectionism is that human cognizers and connectionist systems display important affinities. They share some of same strengths and weaknesses. On the one hand, both are relatively good at tasks involving recognition, transformation, and association, such as in visual recognition and categorization tasks, memory tasks, and complex problem solving (Rowlands 1999, 162). For instance, connectionism offers a superior understanding of cognitive processes that involve learning from experience. It models learning using algorithms (often with a back-propagation training technique) that modify the weights between nodes. Experiments on models have demonstrated cognitive capacities such as motor control, perception, learning, and implicit memory (Piccinini 2012), but also detecting grammatical structure and recognizing faces. On the other hand, both human cognizers and connectionist systems are relatively poor at formal reasoning and performing logical and mathematical calculations, which the classical approach excels at. Thus, connectionist networks model capacities that fit the distinctive cognitive profile of human agents (flexible generalization, adapting to noisy input, etc.) that classical architectures have difficulties with.

At the same time, the growth of cognitive neuropsychology and advances in the study of cognitive impairments related to brain injury have increased the search for computational models that could simulate "injuries," continuing to operate with diminished performance. Classical symbolic cognitive architectures are ineffective in this regard because their models typically either function faultlessly or collapse. In contrast, it is possible to induce "injuries" into connectionist models (by reduction of connections or changes in weights) without the total loss of performance.

Some of these affinities have led to claims that connectionist models are more "biologically plausible" than classical ones. It is, however, difficult to evaluate such claims: we lack a firm comprehension of how cognitive, computational, and neural levels of explanation are linked, and the nodes in connectionist networks fail to reflect the heterogeneity exhibited by neurons (Clark 2014, 87–89; Grush 2002). Still, one may emphasize two aspects that support some modest claim of biological plausibility, at least compared to classicism: First, in contrast to the plastic and resilient nature of brain function, classical computer functioning is particularly susceptible to relatively small errors in both programming and execution. In comparison, both neural structures and connectionist networks often sustain damages without leading to the serious impairment of cognitive capacities. Second, connectionist networks are consistent with the fact that neurons operate at a lower speed of computation than digital computers, unable to execute classicist serial computation quickly enough to match everyday cognitive functioning. This why some have advocated replacing serial computation with the kind of parallel computational architecture that neural networks provide (Rumelhart 1989).

Connectionism may not have achieved a dominant position in the field, but it certainly has managed to cast doubt on the view propagated by Fodor, among others, that classicism constitutes the only robust framework for cognitive science. One key argument against connectionism is that it is unable to accommodate compositional semantics, which is ubiquitous in language and thought. More precisely, the complaint is that connectionism cannot admit of compositional semantics unless connectionist models themselves are implementations of classical architecture (Fodor and Pylyshyn 1988). However, it may be claimed that this argument leads to a false conclusion, implying that connectionist networks are unable to support compositional semantics. Chalmers (1993) argues that connectionist networks can implement Turing machines, which, in turn, can support compositional semantics. Chalmers also puts his finger on something else that is useful to keep in mind with respect to some of the methodological considerations in this book, and in chapter 4 in particular. As he argues: "Behaviorism was very good at explaining conditioning, but it had a problem: it was only good at explaining conditioning. Fodor and Pylyshyn's Classicism is good at explaining compositionality and compositional semantics, but it's not necessarily good at explaining much else. Both conditioning and compositionality are only small aspects of the mind; it seems to be an illusion of perspective that led to behaviorists and Classicists putting so much respective emphasis on them" (Chalmers 1993, 347). The idea that single aspects like compositionality are only parts of the story, and perhaps relatively unimportant for many aspects of our cognitive lives, is closer to Fodor's (2000) more recent considerations, in which he emphasizes that classicism is only a part of the truth about cognition. Although this can be understood as having more or less substantial implications, it arguably raises doubts about the possibility of providing a single and unified theory of cognition.

2.2.3 Mixed Architectures

The previous sketch of classicism and connectionism is admittedly crude, but perhaps comprehensive enough to expose some common theoretical commitments that are important for the next step of our inquiry. Connectionism undeniably poses challenges to classicism, but the changes it engenders, such as the shift from symbols and combinatorially structured representations to distributed ones, leaves unopposed some fundamental principles adopted by classicism. This is in part why classicism and connectionism are best comprehended as two major research programs *within* cognitivism (McLaughlin 2004). Both maintain that the mind exhibits a computer-like functional architecture, comprising a system of functionally organized structures and basic modes of operation. Both aim to identify the functional architecture of the computing system that underlies cognitive capacities and to explicate how they are exercised via systemic operations. Where classicism and connectionism appear to differ the most is in the *kind* of computer the mind is supposed to be and the kind of functional architecture it is supposed to exhibit.

It is thus possible in principle to remain within the boundaries of the overall cognitivist paradigm and to deny that classicism and connectionism are mutually exclusive. For instance, one could deny the often implicitly assumed *architectural monism* (Ramsey 2014), according to which cognitive subsystems must be driven by a single type of representational system and computational architecture. Accepting that the mind cannot be described as a single kind of computer opens the possibility to think of it as an entity that exhibits a hybrid functional architecture, such that cognition displays both symbolic and connectionist features. For instance, given the success of connectionist models in certain areas, perceptual and motor control units may have a connectionist architecture; given the success of classicism in other areas, linguistic and mathematical modules may have a symbolic architecture.

The *integrated connectionist symbol* architecture (Smolensky 1994, 1995) exemplifies an initial attempt to amalgamate classicist and connectionist architectures in a way that is able to explain the systematicity and productivity of cognition. Smolensky and Legendre (2006, 10) highlight hybrid cognitive architecture and the need to overcome what they refer to as a "crisis" in cognitive science, emerging from the observation that the "computational architecture of the brain seems to be connectionist, while the most successful explanation of the mind has depended on a very different architecture, symbolic computation." To the question of what type of computation cognition is, the authors answer that it is both connectionist "brain computation" with parallel processing and symbolic "mind computation" with rule-based transformation of symbol structures: "What makes a mind/brain, what gives rise to human cognition, is a complex dynamical system that is a massively parallel numerical computer at a lower level of formal description—a level closer to the biophysical—and, at the same time, a rule-governed processor of discrete symbolic structures at a higher level of description—a level considerably more abstract than the biophysical, but nonetheless governed by formal laws, and related to the lower level description by precisely defined mathematical mappings" (Smolensky and Legendre 2006, 31). While recognizing that new approaches in machine learning develop machines with a mixed architecture, some have argued that such a mixed account is unable to explain systematicity and productivity without implementing a classical architecture (McLaughlin 2014). However, if a hybrid architecture operates on the assumption that the mind contains different architectures with different representational capabilities, then the systematicity argument loses its force. One can accept that particular parts of cognitive processing involve a kind of representational systematicity that connectionism fails to explain, without assuming that this is a feature that characterizes the entire range of cognitive processes. Discussing architectural pluralism, Ramsey (2014, 254) maintains that "there is no reason to think that explaining representational systematicity is necessary for a viable theory of cognitive processes."

2.3 Cognitivism, Clinical Cognitivism, and Mental Disorder

Cognitivism has led to a tremendous theoretical expansion, and the new outlook on cognition was widely believed to harbor the promise to advance psychiatry and clinical psychology. Many saw the theoretical resources of cognitivism as a platform for rethinking existing explanations of mental disorders and psychotherapeutic methods. But there is a noteworthy dissimilarity between computational modeling approaches that aim to simulate aspects of mental disorders (section 2.3.1) and what we could call *clinical cognitivism*, which seeks to apply some of the main principles of cognitivism to a clinical and therapeutic context (section 2.3.2).

2.3.1 Computational Modeling in Psychiatry

Psychiatrists and cognitive scientists working within the classical computational framework have proposed that aspects of psychopathologies might be constructively addressed in terms of symbol processing. For instance, paranoid thinking was hypothesized to stem from the underlying structure of the patient's rule-based symbol processing. The PARRY program (Colby, Weber, and Hilf 1971; Colby 1981; see also Boden 2006) successfully modeled paranoid response patterns using keywords related to danger and passed the Turing test: its replies were frequently diagnosed as "paranoid" by physicians who didn't know that they were dealing with a computer program. Accordingly, psychiatrists and cognitive scientists working within the connectionist framework proposed that certain cognitive distortions can be seen as resulting from overly strong or weak connections among nodes. Models for processing negative information in depression used expanded computational semantic network models to demonstrate how the overactivation of sad emotion nodes leads to disproportionate activation spreading and ultimately to excessive negative elaboration (Ingram 1984; Tryon 2014).[5] Compared to symbolic models, such connectionist models have the advantage that their distributed processing and storage is vulnerable to damage in a way that is similar to the human brain.

Computational models offer unique possibilities for generating and testing predictions that would not otherwise be feasible or ethically responsible. Contemporary computational approaches in psychiatry, often using principles from computational neuroscience, have the additional advantage of uniting behavioral, cognitive, and physiological aspects (Adams, Huys, and Roiser 2015). Even so, the current landscape only offers models focusing on individual symptoms, and it may seem surprising that there are no computational models of the full range of symptoms of common mental disorders like depression. In the same way, one may wonder why computational schemes fail to address certain crucial aspects like the production of false beliefs (for a discussion, see, e.g., Friston et al. 2014). It is perhaps psychiatry's relative lack of unification, scientific maturity, and validated DSM categories that map onto neurobiological mechanisms that prevents it from benefiting from the full potential of modeling (Huys, Moutoussis, and Williams 2011).

2.3.2 Clinical Cognitivism

Clinical cognitivism largely followed the information-processing model of cognition, regarding sensory and motor realms as functionally separate but linked to a central processing unit, which translates input into output via intermediate connective mechanisms such as representations and cognitive schemata. Inspired by the cognitivist view of minds as analogous to computers, clinical cognitivism pursued a double goal: to achieve a better understanding of psychopathological conditions in terms of *faulty information processing* and to devise tools for initiating positive change through the correction of information processing (Beck and Weishaar 2008; Ingram, Siegle, and Steidtmann 2009).

Incorporation into mainstream psychiatry and clinical psychology proceeded at a very slow pace (Ingram, Siegle, and Steidtmann 2009). For many clinicians, it was not immediately clear how abstract knowledge from cognitive science could account for everyday clinical phenomena, and standard work offered little direct reference to psychiatric disorders and treatment (Stein 1992). Nonetheless, some psychiatrists and clinical psychologists made important attempts to offer a bridge from cognitive science to clinical phenomena, translating cognitivism into cognitive models and therapies mainly based on their clinical practices (Tryon 2014). To illustrate how cognitivist (though not unequivocally computational) ideas have been translated into practice, the following sections explore Aaron Beck's original approach, cognitive therapy,[6] which is enormously influential and presently the first-option treatment for a number of conditions (Hollon, Stewart, and Strunk 2006; Hollon and Beck 2004; Epp and Dobson 2010).[7]

2.3.3 Automatic Thoughts and Cognitive Schemas

Cognitive theory characteristically distinguishes between "schemas" and "automatic thoughts" that arise on the basis of schemas. The concept of a *schema* originated in artificial intelligence (AI) research, where it originally referred to a network of information that could form a database. Schemas are defined as cognitive patterns with "specific rules that govern information processing and behavior" (Beck et al. 1990, 8; Beck 1976). Cognitive theory rightly maintains that experience does not take place in a world that is neutrally disclosed. Rather, we always already are anchored by our *cognitive schemas*—stored bodies of fundamental, background beliefs (Beck and Alford 2009, 255; Williams et al. 1997). The schemas begin developing in childhood and are portrayed as stored bodies of knowledge containing fundamental beliefs, which shape and frame cognition and experience.

One main claim is that individuals suffering from a variety of psychiatric conditions have distorted schemas that lead to emotional disturbances (Ingram, Siegle, and Steidtmann 2009, 84). For example, the schemas of depression-prone individuals are considered dysfunctional because they contain beliefs about self and world that are rigid and unrealistically negative (Epp and Dobson 2010; Blatt 2004). The negatively biased cognitive schemas reflect past experiences and organize new experiences in a biased manner, often resulting in catastrophic interpretations of an occurring situation, which is then rationalized by the schema. In the case of numerous and significant early negative experiences, negatively biased schemas may develop and influence the preferential processing of information.

Over time, these schemas often become unconditionally acknowledged as truths.

Schemas mostly are not readily accessible to consciousness, but they are expressed in "automatic thoughts" (Alford and Beck 1997, 16–17), consciously accessible "self-statements" (Gilson and Freeman 2009) that accompany experience as a kind of "running commentary" (Grant et al. 2010). The thoughts are described as involving "actual words or images that go through a person's mind" (Beck 1995, 16) and consist of mental representations that are speech-like and structurally isomorphic with language. This embraces, but also goes beyond, the classicist claim that representations encode propositional content through a syntactic structure that is language-like.

Practitioners therefore are instructed to convey to their patients that it is of utmost importance to recognize and capture the "precise wording of their automatic thoughts" (Beck and Alford 2009, 311). Cognitive training is supposed to make people able to "catch" the automatic thoughts that follow an event and trigger the emotional response. Schemas become apparent to both the patient and therapist as they identify the "themes" that run through the automatic thoughts and seek to determine the content of maladaptive information-processing strategies. The aim is to capture automatic thoughts that allegedly elicit emotions in the moment and, on that basis, to explore the underlying, general-belief schemas that these thoughts express. In a supplementary step, the attempt is made to change the semantic information that such emotion-eliciting thoughts contain and modify the content of basic schemas. The patient is encouraged to logically challenge and test thoughts against reality and to create substitute representations that help restore positive mood states. In this framework, once the correction is achieved, these schemas can no longer trigger the unpleasant emotional disturbances.

2.3.4 Some Limitations

This is a somewhat simplified account of the theory, but it is comprehensive enough to demonstrate its focus on distorted information processing at the cost of attention to the emotions and the sensorimotor domain. Several chapters will deal with these issues in detail, which is why this chapter will merely point out a few problems with the framework. First, clinical cognitivism leaves open questions about causality. There is abundant evidence that individuals with depression display information-processing biases in memory, self-evaluation, anticipation, and probabilistic inferences to different degrees. They selectively attend to and more readily remember negative material (Peckham, McHugh, and Otto 2010; Dalgleish and Watts 1990), tend to overlook positive material (Winer and Salem 2016), and display compromised reward learning (e.g., Kumar et al. 2008). However, it has been difficult to exclude that these processing biases are not to a significant extent the *products* of depressed affective background states. In that case, the purported causes could instead turn out to be symptoms of the condition.

Second, though psychomotor disturbances can help in discriminating depressive subtypes (Parker 2007), clinical cognitivism only offers a rudimentary view that regards psychomotor retardation as in some way caused by distorted processing and ensuing emotional changes. However, as we shall discuss in later chapters, several findings indicate that significant improvement of the overall condition can be achieved by inducing bodily changes (in gait pattern, posture, relaxation of particular muscles, etc.). These changes not only affect emotional processes, but also can correct cognitive biases—for instance, in the processing of self-referring information. But if this is the case, then clinical cognitivism needs to explain how treating the bodily symptoms of the condition can change the cognitive biases that purportedly cause it.

Third, though vulnerability to depression allegedly depends on individuals' underlying dysfunctional schemas, vulnerable individuals do not seem to remain committed to these schemas once their depression has remitted. Once the patients no longer suffer from the affective disturbances in depression, the relevant negative or dysfunctional beliefs cease to be active or existent. Rather than being enduring characteristics of vulnerable individuals, such beliefs are habitually dependent on background affective states (Teasdale and Barnard 1993). These findings compel acknowledging the role of embodied background emotions in cognitive functioning, which receives support from Antonio Damasio's (1994, 1999, 2003; Lane et al. 2000) work on inconspicuous mood-like background emotions. These are composite expressions of basic and homeostatic and metabolic bodily processes that index momentary parameters of the organism's inner bodily state and express an "image of the body landscape." Background emotions are thus bodily feelings in large part, which do not merely affectively taint experience but also influence cognitive activity.

Phenomenological considerations support the overall idea. For instance, the subtle background emotion "being in danger" can permeate cognitive and evaluative aspects of experience (Ratcliffe 2008). Somebody undergoing this emotion will have a propensity toward holding certain beliefs about herself and the world—for instance, seeing herself as vulnerable and perceiving the world as a place in which her aspirations are unlikely to be realized. Such an emotion thus creates a space of possibilities in which the presence of particular possibilities for action is unambiguous and heightened, while others are experienced as repressed or diminished (Ratcliffe 2010). Importantly, the experiential action possibilities are felt above all as bodily potentialities, such that an altered background emotion simultaneously involves changes in bodily feelings and changes in the space of possibility.

These issues indicate several challenges for clinical cognitivism. It needs to rethink its causal claims in general and to address that psychomotor disturbances seem to have an active causal role in some mental disorders. Moreover, if it is correct that schemas are themselves rooted in background emotions, then clinical cognitivism needs to take into consideration the possibility that the dysfunction in depression may stem from the relationship *between* schemas and background emotions and not from the schemas themselves (Varga 2014). Taken together, the explanation of the dysfunctional response patterns must include bodily and emotional aspects, which clinical cognitivism needs to accommodate.

2.4 Cognitivism and the Separability Hypotheses

The preceding considerations on classicism, connectionism, and hybrid approaches are helpful to uncover major points of disagreement, but also underlie a general theoretical commitment to representationalism, combined with the claim that cognition is computational. Although some connectionist and hybrid models have distanced themselves from the view that minds are logical reasoning devices, their development of a different vision of computational structure remains committed to the classicist conceptualization of cognition in terms of input-output mappings (Clark 1997, 80; Kolak 2006). For all cognitivist approaches, the boundaries of cognition drawn in cognitive science are identical to the boundaries of computation drawn in computer science (Shapiro 2011), where the relevant processes occur in between (1) receiving the input and (2) displaying the output. Just as attending to what happens between (1) and (2) can suffice for the needs of the computer scientist, the cognitive scientist can limit an inquiry to processes within the head, without having to consider other features of the organism or the world it resides in. Thus, according to this shared theoretical commitment, the body and world are *separable* (in a yet-to-be-determined sense) from cognition.

How is this separability thesis to be understood? Cognitivism acknowledges that human minds are embodied, but only in an explanatorily weak sense. Cognitivism acknowledges that cognition could not exist without a functioning bodily machinery (*existence-dependence*) and that cognition requires stable contact with the body and the world, such that it can constantly be informed by our bodily interactions with our environment (*input-dependence*). In this sense, cognition depends on the body and its ability to manipulate perceptual input—for instance, by moving toward or away from the stimulus. While acknowledging that the body can thusly shape the input, cognitivism maintains that the brain is where all cognitive processing occurs and that the body is not something that one is required to investigate to understand cognition. In other words, it denies that cognition is dependent on the body in the sense of *nature-dependence*.

It is this sense of dependence that is most relevant for the purposes of clarifying what separability means. The separability thesis can be formulated as two theses, one about the location of cognitive processes and the other about how these processes are best studied and understood:

(1) *Ontological separability:* Although standing in different relations to events, states, and processes in the cognizer's body and environment, cognitive states and processes (involving formal operations on representations, connectionist network activation, or mixed architectures) are located exclusively in the head of the cognizer.

(2) *Epistemological separability:* Although these events, states, and processes may play an important causal role in shaping and facilitation cognition, we can provide a full understanding of cognitive processes by studying exclusively what is occurring between bodily input and output, which, per thesis (1), are located inside the head of the cognizer.

The principles that constitute the separability thesis lead to what has been called a *disembodied* picture of the mind. Corresponding to these principles, the idea that cognition can be fully explored in abstraction from the non-neural body is reflected in the cognitivist idea that the appropriate method for psychology is "methodological solipsism" (Fodor 1981). The view is that the organism's sensorimotor systems merely provide sensory input and permit behavioral output that can be detached from autonomous mental processes.

2.5 Embodied Cognition and the Inseparability Hypothesis

Over the last two decades, cognitive science has witnessed the emergence of a new view that opposes the cognitivist picture, rejects at least one of the two principles comprising the separability thesis, and envisions a nondisembodied view of cognition (see, e.g., Varela, Thompson, and Rosch 1991; van Gelder 1995; Thelen and Smith 1994; Clark 1997, 1999, 2008; Brooks 1999; Rowlands 1999, 2010; Lakoff and Johnson 1999; Gallagher 2005; Wheeler 2005; Gibbs 2006; see also Anderson, Richardson, and Chemero 2012). The view is that behavior, the particular structure of the body, and environmental resources are far more deeply implicated in cognitive activity than cognitivism suggests, such that "if we are to understand mind as the locus of intelligence, we cannot follow Descartes in regarding it as separable in principle from the body and the world" (Haugeland 2000, 236). As Haugeland (2000, 237) notes, the mind is "not incidentally but intimately embodied and intimately embedded in its world."

This change in perspective has propelled the emergence of the EC research program, which positions itself against one or both separability theses and comprehends cognition not as the product of the brain, but as the activity of a more comprehensive system, in which brain, body, and environmental features dynamically interact. Instead of thinking of cognition as functionally disconnected from sensorimotor systems, many advocates of EC think that cognition is grounded in the sensorimotor system of an agent situated in an environment in many cases, best explicated in terms of a dynamic interaction between neural and bodily processes in interplay with the environment (Foglia and Wilson 2013). Moreover, EC is united in stressing that perception and cognition are primarily used for action carried out in real time and in denying that cognition is "pure" problem solving, which, as Andy Clark (1997, xii) puts it, "invites us to abstract away from the very body and the very world in which our brains evolved to guide us."

Although EC involves several relatively heterogeneous fields of research that in some cases operate with slightly dissimilar concepts, claims, commitments, and research foci (Wilson 2002), we may nevertheless attempt to characterize a basic commitment shared by most EC accounts, which is that numerous aspects of an agent's cognition depend (nature-dependence) upon features of the agent's non-neural body and environment. EC accounts probe various versions and implications of this basic claim. This formulation is broad enough to include accounts focusing on how cognitive processes of individual agents may be grounded in sensorimotor systems, as well as accounts focusing on the situated nature of cognition, which may include offloading certain cognitive tasks onto the environment itself or into bodily processes that become proper parts of cognition. In contrast to cognitivism, EC can be described in terms of inseparability theses:

(1) *Ontological inseparability:* Cognitive states and processes are not located exclusively in the head of the cognizer. Instead, the material realizers of cognitive processes can sometimes include parts of the body and perhaps even resources beyond the boundary of the body. Some might deny that cognition is anything like the problem-solving steps specified by a computer program, whereas others may accept that the operations that realize cognition are computational in some sense (involving formal operations on representations, connectionist network activation, or mixed architectures) but maintain that the hardware itself can extend beyond the brain.

(2) *Epistemological inseparability:* The organism's body and sensorimotor systems do not merely deliver sensory input and enable behavioral output, but also actively participate in carrying out cognitive activity. We cannot provide a full understanding of cognitive processes by studying exclusively what is occurring between bodily input and output located inside the head of the cognizer. Human cognition depends (nature-dependence) on having a body with distinct sensorimotor capacities and being embedded in an environment that supports them.

These theses are in many ways interconnected, but they can be held and defended separately. The truth of thesis (2), provided one holds an additional assumption of explanatory success in justifying metaphysical claims, might be taken to support thesis (1). Likewise, the truth of thesis (1) would lend support to thesis (2): if cognition cannot be separated from the body that supports it, then an explanation entirely by appeal to internal, representational states will not be adequate.

These two inseparability hypotheses only provide a brief sketch of some of the commitments of the EC research program. Although one may be tempted to evaluate EC's merits by drawing parallels to cognitivism, one should keep in mind that EC is in its early stages and lacks characteristics that more mature programs display. It cannot be unified into a single theory that is able to deal with the entire range of empirical material. First, EC's domain is not as clearly demarcated as cognitivism's. Whereas the latter focuses on standard cognitive processes (perception, memory, attention, language, problem solving, learning), many EC authors study the role of the body and its movements in assisting and constraining memory tasks and spatial reasoning, specific features of subsumption architecture in robotics, the development of walking in toddlers, and neuroscientific evidence that higher-level cognition involves motor simulations. Second, though the main ontological commitments of cognitivism are relatively clear, EC operates with different sets of commitments regarding mental representations and computation, some of which are, in effect, consistent with cognitivism. Third, to a large extent following from the first two points, the lack of sharp domain demarcations and theoretical commitments is prone to result in dissimilar methodological commitments. For these reasons, it is not clear how to evaluate the claim that EC is going to replace or supersede standard cognitive science.

2.6 The Roots of Embodied Cognition

The previous sections have introduced EC as a reaction to cognitivism in cognitive science, in part because it is beneficial to approach a somewhat inhomogeneous research program by highlighting the views it reacted against. But in part because of the inhomogeneous nature of EC, an introduction is not complete without a sketch of the most important philosophical and psychological roots and sources of inspiration that were instrumental in the development of EC. These sources span the phenomenological tradition, ecological psychology, dynamic systems theory, and robotics.

The phenomenological tradition has served as a significant source of inspiration for EC (see Gallagher 2014; Gallagher and Zahavi 2010; Varga 2018a), but a detailed charting of its particular influences would lead us off topic. In particular, EC authors have drawn on Merleau-Ponty's work, acknowledging its merits in taking decisive steps toward rejecting the attempt to localize semantics in a symbolic realm functionally detached from sensorimotor systems and toward understanding the mind as intertwined with body and world. Prominent EC theorists develop impulses from Merleau-Ponty in different ways but subscribe to the phenomenological view that the cognitive organization of human beings echoes their bodily immersion in the world.[8] Some of them more implicitly (Varela, Thompson, and Rosch 1991, 172–179; Clark 1997), and others explicitly, endorse the idea that such embodied immersion leads to a characteristic "circular causation" between agent and world that is crucial for understanding cognition.

On the less radical end of the scale, some draw inspiration from the phenomenological tradition to highlight embodiment as a constraining factor on cognition. Inspired by Merleau-Ponty's acknowledgement of the fact that "our bodily experience is the primal basis for everything we can mean, think, know, and communicate," Lakoff and Johnson (1999, xi) highlight abstract concepts and metaphors as grounded in particular domains of bodily experience and elucidate how embodiment enables and constrains cognitive processes.[9]

On the radical end of the scale, Varela, Thompson, and Rosch (1991) explicitly acknowledge their work as "a modern continuation of a program of research founded over a generation ago by ... Merleau-Ponty" (xv) and attempt to redirect the focus of cognitive scientific investigation from representations to the sensorimotor interactions between agent and world, conceiving of cognition as "embodied action" (xx). They argue that cognition is the "history of structural coupling that brings forth a world" and that cognitive structures "emerge from the recurrent sensorimotor patterns" (173). Cognitive processes work "through a network consisting of multiple levels of interconnected, sensorimotor subnetworks" (206), while human embodiment, as a constraining factor on human cognition, reflects only one of many possible evolutionary pathways (213–214). Two basic tenets of their account are that "cognition depends upon the kinds of experience that come from having a body with various sensorimotor capacities" and that "these individual sensorimotor capacities are themselves embedded in a more encompassing biological, psychological, and cultural context" (173). Perception and action are not merely contingently connected but have evolved together.

Somewhere between these two accounts, we find Clark's pioneering work that also draws on ideas from the phenomenological tradition. He takes seriously Heidegger's (1962) point about our being active participants (as opposed to detached, passive observers) in the world, as well as his emphasis on how our practical dealings often involve functional couplings instead of detached representations. Although the phenomenological idea that skillful practical engagement with the "affordances" of the environment is crucial for cognitive activity to a certain extent informs Clark's criticism of the idea of "action-neutral kinds of internal representation," Clark (1997, 148) does not support the strong thesis according to which cognitive processes are best comprehended in noncomputational terms. The reason for this is Clark's conviction that new data and conclusions about the tight coupling with the environment can be incorporated into an adequately modified computational framework (Clark 2014, ch. 7).[10]

Finally, somewhat parallel to the development of Merleau-Ponty's later work on the "reciprocity" of perceiver and environment, Gibson (1979) made significant advances in the field of ecological psychology, developing his notions of *affordances*, which are opportunities of interaction that environmental features possess in relation to an organism's sensorimotor abilities. Without dwelling too much on the similarities, it is relevant to highlight two shared commitments. The first focuses on how perception does not merely consist of inferences performed on sensory representations but can put us into direct contact with the world, unmediated by representations. Gibson questions a basic assumption about the nature of the challenge that perceptual systems confront. Whereas cognitivist accounts, such as the computational vision program (Marr 1982), assume that the visual system faces the task of having to create a comprehensive three-dimensional impression from the poor information that the two-dimensional retina image offers, Gibson argues that due to the organism's being already attuned to relevant affordances, relatively simple mechanisms use information made accessible in the environment and are thus able to carry out tasks that are usually assumed to necessitate complex processing of information.

The second commitment opposes a stark dichotomy between action and perception, emphasizing that "we must perceive in order to move, but we must also move in order to perceive" (Gibson 1979, 223) and that perceptual systems have coevolved with motor capacities and particular environmental structures and "niches" that are partly results of human impact. Roughly put, perception is for the generation and control of action, and much action is for perception or cognition. The perceptual systems of organisms are organized not to create precise internal representations of the environment to operate on, but primarily to enable action by picking up goal-related affordances. Accordingly, relatively uncomplicated nonrepresentational mechanisms, through which the body interacts with a structured environment that embeds it, are able to solve many cognitive tasks without having to "translate" sensorial information into representations, leading to operations, leading to action. The concept of affordance directly links perception, action, and the environment and lends itself to a study of perception that is not conducted as detached from the study of behavior.

Some philosophers early on highlighted important parallels between the rise of the idea of EC and the emergence of a richer conception of intelligence in robotics (Clark 1997). EC has also drawn on a particular strand of work in computational intelligence that became known as the *embodied approach to robotics*. Before the 1990s, work in robotics standardly adopted a general cognitivist view, in which sensory input is analyzed and transformed into abstract amodal symbols that represent a state in the world. Learning or decision-making processes in the robot then use these states to choose particular actions, executed by actuators, which alter the state of affairs of the robot's environment, giving rise to new sensory input. The processing of information occurs in a separate realm, and the flow of information is unidirectional: from the environment to sensory input, decision making, and action, then back to the environment (Hoffman 2012, 2). This classic understanding of cognitive tasks as representation-intensive processes was challenged by a new "subsumption architecture" (Brooks 1991), which breaks down complex behaviors into series of simpler behavioral arrangements. These systems are not integrated (in the sense that they form a coherent representation of the world), and each of them is designed to meet a specific goal and respond to features of the environment, organized such that a layer's goal is "subsumed" by the one above it. Although the architecture involves control layers and simple interactions between self-contained subsystems, the system operates without a central instance that controls the activities of the different layers.

A favored example in the literature is that of the mobile robot that navigates the environment without bumping into objects. The robot "senses objects in its immediate vicinity and moves away from them, halting if it senses something in its path. It is still necessary to build this system by decomposing it into parts, but there need be no clear distinction between a 'perception subsystem,' a 'central system' and an 'action system.' In fact, there may well be two independent channels connecting sensing to action (one for initiating motion, and one for emergency halts), so there is no single place where 'perception' delivers a representation of the world in the traditional sense" (Brooks 1991, 147). For EC proponents, the success of these robots illustrates that cognitive processing is not always required and can be replaced by coupling perception-action loops to the relevant parts of the environment. There thus can be intelligent behavior that is "world-driven" and does not heavily rely on representations. Whereas "cognitivist robotics" would aim at a sequence of independent processes consisting of perception, forming representations, operating on them to plan action, and issuing a motor command, this robot's layers are synchronized not by internal computation, but by the environment itself.

One important theoretical upshot for EC is that computational intelligence does not necessarily require complex internal algorithms and representations but can use relatively simple bottom-up controls governed by interaction with the world. Moreover, minds may not be devices primarily for abstract thinking in the traditional sense, and intelligence is not reducible to symbol processing occurring between perceptual modules and independent motor devices but denotes instead the intertwined operation of perception-action loops. Acknowledging this function of the body and the environment, "it is fitting to cease viewing their physical presence as a noisy filter over pseudoperfect information. Instead, and especially when interacting with humans, robots should make use of their embodiment, as the human brain does with its own physical circumstance" (Hoffman 2012, 11).

Finally, EC has received important inputs from the emerging dynamic systems approach to cognition, which is developed from the analysis of complex physical systems and applies the tools of dynamical systems theory to the study of cognition (Thelen and Smith 1994; Beer 1995, 2003, 2014; van Gelder 1998a; van Gelder and Port 1995). The mathematical principles that characterize dynamic systems are continuous with those of connectionist networks (Horgan and Tienson 1992a; Port 2002). Much like the way ecological psychology comprehends perception-action loops as changes in a system that includes features of the organism and the environment, dynamical systems theory allows that cognitive performance might include variables that are external to the agent. Moreover, its two major theses bear noteworthy resemblances to the ontological and epistemological inseparability theses in EC, including assumptions about how they could support each other. As van Gelder (1998a, 619) maintains, "The *nature* hypothesis is a claim about the nature of cognitive agents themselves; it specifies what they *are* (i.e., dynamical systems). The *knowledge* hypothesis is a claim about cognitive science: namely, that we can and should understand cognition dynamically. Obviously, these are closely related; the best evidence for the former would be the truth of the latter."

Consider, for instance, moving your hand to grab the cup on the table. Cognitivism would maintain that in possession of accurate representations of the environment, you perform a series of computations (about weight, distance, gravity, etc.) to coordinate muscle contractions, much like your self-driving car calculates speed, distance, engine power, and so on in order to park. To illustrate an alternative way of thinking about such tasks, van Gelder (1998a) considers an example by Kelso (1995), which involves the coordination of the wagging of index fingers. At low speeds, two coordination patterns (inphase and antiphase) characterize the movement, but at higher speeds only inphase patterns remain stable. Dynamical systems theory explains this by positing a single evolving high-level variable, and it maintains that "coordination is best thought of as explained not as masterminded by a digital computer sending symbolic instructions at just the right time, but as an emergent property of a non-linear dynamical system that self-organizes around instabilities" (van Gelder 1998a, 616). But in that case, the key to stabilizing adaptive coordinated patterns of behavior is not the accurate representation of the environment, but rather the continuous bodily engagement with it. This shift in emphasis helps pose an intriguing question to proponents of cognitivism: "If the environmental stimulation impinging on our sensory systems is almost always partially overlapping in space and continuous through time, why would the mind work in a staccato fashion of entertaining one discrete stable non-overlapping representational state for a period of time, and then instantaneously flipping to entertain a different discrete stable non-overlapping representational state for another period of time?" (Spivey 2007, 8). Proponents of dynamic systems theory maintain that the emphasis on continuous bodily engagement is a better fit with the structure of environmental stimulation and reduces the need to posit internal representations and computational structures as mediators between the world and internal processing to explain cognition. Of course, such a view requires rethinking the possibility of misrepresentation—that is, instances in which an internal representation fails to track the right external referent. But the reply by proponents of dynamic systems theory is fairly simple. Occurrences that might at first look like misrepresentation often are better considered not as mistakes in representation, but as deficiencies of agent-environment engagement and synchronization (Beer 2003, 238).

2.7 Conclusion

The "cognitive revolution" rehabilitated mental phenomena as suitable subject matters for scientific psychology and established itself as offering a new framework for investigating and understanding psychological phenomena. When cognitive science emerged as a discipline, its theoretical framework was to a large extent constituted by a representational theory of mind, combined with the thesis that cognition is computational. This framework is intelligible as a response to behaviorism and the development of artificial intelligence's emphasis on logic and computer technology. Furthermore, it has proved productive. It enabled the rise of a relatively unified interdisciplinary approach that explores cognition by offering an ever-better comprehension of the nature of representational structures and of the particular causal processes that execute computational operations on them. Also, it enabled taking decisive steps toward naturalizing the mind, or at least sidestepping thorny metaphysical questions about the ontological status of mental states within the physical order of the material world.

This chapter described the cognitivist paradigm as incorporating three partially competing approaches (classicism, connectionism, and mixed architecture), which all postulate causally active, internal processing procedures and regard the mind as a computer that processes information from the environment to perform intelligent actions. With the limited amount of space dedicated to describing the main tenets of cognitivism, it may appear as more unified than it actually is. There are additional disagreements, beyond the ones linked to the range of forms that representation and computation can take and the type of computer architecture that captures the essence of intelligent behavior. However, they share a commitment to the principles of ontological and epistemological separability.

EC, meanwhile, draws inspiration from several different sources (phenomenology, ecological psychology, dynamical systems theory, and robotics) and is in many ways still in its early stages as a research program. It lacks the characteristics of a well-defined theoretical approach and cannot be unified into a single theory that is able to deal with the entire range of relevant empirical findings. Although the lack of homogeneity is not a decisive problem from the vantage point of this book, the future status of EC will depend on clarifying whether it should be split into several enterprises, some of which are perhaps best understood as useful additions and corrections to the cognitivist approach to cognitive science, or whether it can develop into a unified area of research that offers an alternative to cognitivism. Such a clarification will likely provide answers to critics who acknowledge that EC reflects interesting emerging trends in experimental design and cognitive modeling but maintain that its theoretical import is "substantially oversold" (Rupert 2009, 242).

Perhaps this is correct and EC should be seen in a more modest light, not as overly revolutionary but as shifting focus to less representation-heavy processes, which does not amount to a wholesale rejection of the notion of representation and computational theory. Chapter 4 will propose a pluralist metaframework that points in a similar direction. But whether the interpretation of EC is as a revolutionary overthrowing of an orthodox paradigm or a shift in focus toward a more comprehensive picture of cognitive processes that considers the body is not decisive for the purposes of this book. First, the investigation in this book focuses on particular themes in EC, guided by their relevance for the ASC framework and the explanation of psychiatric phenomena. Second, though chapter 3 will identify some relevant commitments that bring some unity to EC, chapter 4 will discuss a pluralist approach to understanding cognition, resisting the view that the EC label is best replaced by several labels that more precisely reveal the differences in projects that have been lumped together under a single label.

3 The Scaffolded Mind

3.1 Epistemological and Ontological Inseparability

Whether the developments that have guided a theoretical reorientation away from cogitivism constitute a genuine paradigm shift is uncertain. Nonetheless, EC, as a conceptual umbrella for a loose-knit group of research endeavors, has definitively influenced the trajectory of much theorizing and experimenting in cognitive science. Labels like *embodied, embedded, extended,* and *enacted* cognition endorse *epistemological inseparability* (the organism's body and sensorimotor systems actively participate in the execution of cognition, such that we cannot provide a full understanding of cognition by studying exclusively processes occurring inside the head of the cognizer). Most of them share a commitment to regarding the mind as the explanatory basis for behavior and understand mental states as playing a causal role in governing it, but they significantly differ in their claims about the extent to which mind and world are entangled (see, e.g., Wilson 2002; Wilson and Golonka 2013; Ziemke 2003). For instance, opinions are divided on the question of *ontological inseparability* (whether the material realizers of cognitive processes can sometimes include parts of the body and the environment), on what role, if any, mental representations should have in explaining cognition, on how to justify the choice of grain level for functional analysis (see a detailed discussion in chapter 4), and on whether thinking that cognition is profoundly embodied is consistent with the functionalist abstraction from biological specifics. In addition, some use a notion of embodied cognition in a way that includes what others call *situated* or *embedded* cognition, some consider extended cognition a distinct class of situated cognition (Robbins and Aydede 2009, 3), some maintain that "work in situated cognition is best viewed as an ongoing series of investigations into cognitive extensions" (Wilson and Clark 2009), and some regard extended cognition as a form of niche construction (Sterelny 2010).[1]

In light of these issues, the distinctions in the literature are best understood in a pragmatic manner, such that some of them will be useful for those who aim to understand how embodied cognition differs from nonembodied cognition, whereas others will be useful to those who wish to understand how embodied cognition differs from embodied noncognition (Shapiro 2014). Rowlands (2015, 311) is probably right to maintain that "claims that mental processes are embodied, enacted or extended have become interpreted in so many different ways that it is, perhaps, no longer advisable to define one's view in terms of them."

Correspondingly, this book will generally refrain from using these labels for carving out the theoretical framework that it proposes. Instead, this chapter will use the notion of *scaffolding* to present and motivate a framework that integrates a medley of positions usually subsumed under the label of *embodied cognition*, along with other accounts that highlight the various manners in which cognitive processes nontrivially depend on the body and the environment. Moreover, this chapter will offer a taxonomy of different forms of scaffolding. There are several ways to taxonomize scaffoldings, such as by ontogenetic and phylogenetic aspects, by cognitive domains, or by the material constituents of the scaffoldings (for additional possibilities, see Sutton 2015). This chapter will propose a taxonomy that is, once again, guided by pragmatic and explanatory considerations regarding common symptoms in mental disorders. Thus, instead of focusing on a particular cognitive domain or function, the chapter will distinguish between two forms of *intrasomatic scaffolding* (simple and complex) and a specific form of *extrasomatic scaffolding*.

3.2 Scaffolding and Active Scaffolding

Although the notion of scaffolding is familiar from developmental psychology, the philosophy of biology, and cognitive science (Clark 1997; Williams, Huang, and Bargh 2009; Sterelny 2010, 2012; Sutton 2015; Wimsatt 2014; Colombetti and Krueger 2015), this book deploys it in a deliberately broad manner to offer the kind of conceptual flexibility that is suitable for our aims. First, scaffolding can refer to a structure with one or more platforms that substantially support an activity or skill, whether artificially created for that activity (erected for workers to repair or construct a building) or simply refunctioned to support it. It thus can capture features that arise through different phylogenetic or developmental processes to support and augment cognition. Second, scaffolding is neutral on the material that the relevant structure is composed of. It can include neural, physical, cognitive, and social parts and thus forms of enduring or temporary support by, for instance, tools, sensorimotor circuits, or other human beings. Third, scaffoldings can become redundant at a certain stage of a process, but they can also become permanent structures that form proper parts of the entity or process that they support. Fourth, it is neutral in the sense that it does not entail particular a priori assumptions about the nature of cognition. This flexibility will help sidestep unproductive debates in favor of explanatorily helpful distinctions, but it also facilitates crucial integrative efforts.

The notion of scaffolding draws on Kim Sterelny's work on cognitive niche construction, which claims to offer a better alternative to current approaches like the extended cognition thesis. The point is not that the main thesis is false, but that it "obscures the varieties of ways in which we engineer the environment and rely on external resources to enhance our cognitive competences" (Sterelny 2010, 473).[2] Building on this line of thought, this chapter regards a wide range of cognitive processes that are embodied and embedded in different ways as particular cases of the broader phenomenon of scaffolded cognition. The concept of scaffolding will assist in constructing an integrative framework that captures a broader range of phenomena and several more or less integrated cognitive systems that include not only the use of resources that human agents are coupled with (enactivism) or that are highly trusted (extended cognition). It will thus link several accounts that subscribe to epistemological inseparability and generate continuity between them, with the aim to explore aspects of mental disorders and new possibilities for therapeutic intervention.[3]

There are elements of each of the considered views that are not endorsed, and the clarification of such divergences will ensue in the chapters that discuss possible applications. Moreover, because the term scaffolding includes a rather large number of relationships of dependence, it is necessary to exclude trivial ones. This is in part achieved by introducing the notion of *active scaffolding* to delineate a particular relationship characterized by nature-dependence and difference-making between the scaffolding and what is scaffolded. As to the first, the relationship of dependence referred to as *nature-dependence* means that the scaffolded (e.g., a cognitive process or an emotion) depends on the scaffold (e.g., the body) not merely for its existence, but also for its nature or character. As to the second, *difference-making* aims to capture that it is possible to manipulate the scaffolded by manipulating the scaffold. At least for present purposes, this helps exclude trivial forms of dependence and difference-making. The exact nature of the metaphysical relationship will be clarified in chapters 4 and 5.

But even without these important details in place, we see that speaking of *actively scaffolded cognition* (ASC) offers a number of advantages for our present purposes. First, we can consider scaffolding structures that serve as support for both cognitive and emotional capabilities,[4] which is particularly useful for investigations of mental disorders. Second, in contrast to extended cognition and niche construction models, ASC includes structures and processes that are not offloaded to the environment but constitute intrasomatic scaffoldings—for instance, by reusing neural and sensorimotor structures for higher-order cognitive processes. Although such structures characterize "the kind of brain that is adaptive to highly structured sociocultural niches that are bequeathed to future generations" (Menary 2014, 291), ASC allows for investigating how intra- and extrasomatic scaffolding relationships can operate simultaneously to support emotions and cognition. Third, the flexibility of ASC not only will help to tailor a taxonomy of forms of scaffolding for the specific task but also will enable it to assign a place for a distinct type of socially scaffolded cognition (*intersomatic scaffolding*) that has not been discussed in the literature. ASC is consequently the genus, and intra- and extrasomatic scaffolding are its classes.

3.2.1 Simple Scaffolding: Botox for Depression?

The work of Antonio Damasio has been significant for uncovering how certain affectively laden bodily states are vital for cognition (Damasio 1994). His work highlights that specific bodily feedback—for instance, in the form of affectively laden "gut reactions" (e.g., disgust, nausea)—guides and constrains a range of cognitive processes. Thus, beyond merely offering life support and sensory input, Damasio maintains that the body supplies "a content that is part and parcel of the workings of the normal mind" (Damasio 1994, 226). He starts with two observations: First, individuals with damage to the ventromedial sector have a severely diminished ability to navigate social conventions and to make adaptive decisions. Second, though learning, memory, language, and attention abilities remain largely preserved, these patients exhibit an abnormal affective profile and fail to engage their emotional reactions to situations and events. These two observations led to the development of the somatic marker hypothesis. Somatic markers denote a special instance of feelings, consisting of interoceptive and proprioceptive feedback from the body. They are processed in the emotion circuitry of the brain[5] and help regulate decision-making in complex situations. Through learning processes, certain bodily feelings have been connected to certain scenarios and their impact on the agent. For example, a negative somatic marker conveys knowledge retained about the previous impact of certain circumstances and can—if properly integrated into decision-making—function as a warning against a potentially disadvantageous choice. Somatic markers can operate outside awareness, but in some cases—for instance, when we experience a "gut feeling"—they reach conscious awareness.

For illustrative purposes, consider a series of studies using the *Iowa gambling task*, in which participants were requested to select cards from four decks (Bechara et al. 1997, 2005). To begin, participants were not aware that some decks were advantageous and others disadvantageous. Healthy subjects started displaying beneficial decision-making before gaining conscious knowledge about the decks, as well as more substantial anticipatory skin-conductance responses just before picking a card from a disadvantageous deck. In healthy subjects, successful performance was found to be correlated with the magnitude of somatic marker signals, measured by anticipatory skin-conductance responses.[6] In contrast, individuals with damage to the ventromedial sector lacked these signals, a lack that was linked to their reduced performance.

The account of somatic marker signals aiding cognitive processes provides helpful guidance in illustrating what "simple scaffolding" aims to capture. But to illustrate the idea more accurately, consider some surprising effects of botulinum toxin type A (Botox). Botox is commonly injected into the facial muscles for temporary paralysis that reduces the appearance of wrinkles. Somewhat surprisingly, it turns out that these injections also produce significant improvement in depressive symptoms (Parsaik et al. 2016). Simple scaffolding provides a tentative model for explanation. Typical emotional states in depression—such as anger, fear, and sadness—involve the contraction of the corrugator muscle in the glabellar region. The temporary paralysis of these muscles interrupts the normal circuitry and diminishes amygdala responses to negative stimuli (Magid et al. 2016; Finzi and Rosenthal 2014). The relevant mental processes are thus actively scaffolded onto sensorimotor circuits such that bodily activation in the form of muscle activity not only supports but also affects them, in ways we will explore in chapter 5.

With this general idea in mind, consider a different case. Studies on motor pathologies and language processing confirm a strong link between the motor system and language comprehension. For instance, patients with Parkinson's disease often display impaired language processing as a result of reduced action possibilities (Fernandino et al. 2013; Cardona et al. 2014). Looking closer at this link, we see that language comprehension partially activates motor structures, such that when participants read a number of action words, the same structures in the motor and premotor cortex are activated as when they initiate the actual movements corresponding to the verbs presented (Hauk, Johnsrude, and Pulvermüller 2004; Pulvermüller and Fadiga 2010; Binder and Desai 2011). Importantly, the activation does not only involve brain structures. In a recent study, participants were asked to evaluate state- and action-verb sentences, during which the experimenters measured their facial muscular activation using electromyography. When reading the sentences, participants exhibited emotionally congruent facial muscular activation both when reading action verbs directly linked to facial expressions (e.g., "Mario smiles") and when reading state verbs describing emotions (e.g., "Mario enjoys"). The corresponding facial muscle activations were differentiated in the sense that state-verb sentences received higher valence ratings compared to action-verb sentences (Fino et al. 2016; see also Foroni and Semin 2013).

Corresponding to the view defining ASC—namely, that active scaffolding designates a particular relationship characterized by *nature-dependence* and *difference-making*—simple scaffolding hypothesizes that if particular cognitive activity is scaffolded onto a sensorimotor circuit, then manipulating the latter in the appropriate manner will affect the former.[7] If, for instance, comprehending an emotion requires the activation of the bodily state that is associated with that emotion, then congruent emotions should facilitate comprehension, whereas incongruent ones should inhibit it. In a study by Havas, Glenberg, and Rinck (2007), subjects held a pen either between the teeth, which stimulates smiling, or between the lips, which inhibits smiling. They were given the task to assess whether sentences presented to them communicated pleasant or unpleasant events. The study showed that pleasant-event sentences were read faster in the smile-inducing condition and slower in the smile-inhibiting condition. Correspondingly, reading times for sentences describing unpleasant events were faster in the smile-inhibiting condition and slower in the smile-inducing condition. Similarly, studies investigating the effects of the temporary paralysis of facial muscles found that paralysis interrupted the processing of emotional facial expressions and sentences but not of emotion-neutral material (Baumeister, Papa, and Foroni 2016).

Circling back to the initial Botox example, there is thus evidence that bodily activation such as facial muscle activity not only plays a supporting role in a higher-order cognitive function but also can causally affect it. This opens possibilities for developing interventions into cognition and emotion via the body to improve the symptoms of mental disorders, while taking into account that there are cultural differences in muscle-activation patterns for the same emotions (Elfenbein et al. 2007). In fact, some simple scaffolding structures are already being exploited for therapeutic purposes. As mentioned, the temporal paralysis of the corrugator muscle in the glabellar region interrupts the circuitry of emotional states in depression and reduces amygdala responses to negative stimuli (Finzi and Rosenthal 2014; Wollmer et al. 2012; Wollmer 2014; Magid et al. 2016). A recent metastudy of exclusively randomized controlled trials evaluating the efficacy of Botox on depression found that inducing temporary paralysis produces significant improvements in depressive symptoms (Parsaik et al. 2016).

3.2.2 Complex Scaffolding: Tylenol for Social Pain?

Although the category of simple scaffolding demonstrates that cognition is more deeply dependent on emotions and the body than previously supposed, other approaches address the manner in which the body scaffolds cognition and language. This view to some extent opposes traditional functionalism, which maintains that the nature of physical material that implements cognition is irrelevant: cognition is a functional kind, in principle realizable by many different physical substrates (Putnam 1975). If mental states are understood in terms of functional relations between inputs, other states, and outputs, then any material that allows the instantiation of the appropriate functional organization will be able to support the right functional properties. Consequently, on the functionalist account, the type of bodies we have, the nature of the physical material implementing cognitive processes, has no bearing on the concepts and cognitive routines we develop.

Opposing this view, a group of accounts maintaining that cognition is embodied promote a version of the inseparability thesis. They deny not only that cognition is independent of its physical vehicle but also that a mind just like ours could in principle inhabit a very different body with very different properties. Instead, the claim is that cognitive activity reflects the physical bodies that support it, such that an optimally firm body of knowledge about the details of an organism's cognition will be able to predict facts about its body. Take, for instance, human vision and audition. They both involve moving the visual sensor equipment, allowing us to acquire information about the relative distances to entities that our perception presents. In the case of vision, the movements of our heads do not simply assist visual processes. Rather, visual processes include features of the human body and would be incomplete (or at least of a different type) without this specific contribution. If it is true that human perceptual processes include putting into effect certain bodily features and skills, then it is warranted to think that it requires having bodies that exhibit those skills (Noë 2004).

Subscribing to the idea that cognitive activity somehow reflects the specifics of the body, a fertile line of inquiry investigates the function of the body in the acquisition and organization of concepts (Barsalou 1999, 2008; Anderson 2003; Niedenthal et al. 2005; Glenberg et al. 2005; Lakoff and Johnson 1999; Lakoff 2014; Andres, Olivier, and Badets 2008). The main claim is that some basic concepts are scaffolded onto the human agent's sensorimotor capacities, which also partially explains how they acquire their meaning. For instance, an everyday concept like "chair" is not definable in terms of a set of objective features, but denotes an object that affords sitting on it. Under certain circumstances it is intelligible to refer to both a leather recliner and an empty box as chairs, even though they lack objective similarities (Anderson 2003). However, whatever propositional knowledge we acquire of the fact that a chair is for sitting, it must be combined with some type of a bodily know-how, normally acquired by mastering the skill of sitting. The general idea, then, is that concepts are scaffolded onto the body and its practical bodily engagement with the world.

This leads to questions about how the sensorimotor domain, through bodily scaffolding, affects higher-order cognition. There are several possibilities here, depending on how scaffolding is conceptualized. Cognition can be scaffolded onto the body as follows:

(1) Such that it involves simulations of bodily experiences. Whenever individuals represent an object in their thinking, there is a partial reenactment of the bodily states linked to the experience of that object (Barsalou 2008, 2010; Barsalou et al. 2005).

(2) Such that it draws on neural resources that primarily have motor functions. For instance, in language comprehension, "sentences are understood by simulating sentence content using neural systems ordinarily used for perception, action, and emotion" (Glenberg 2010, 589). The comprehension of a sentence involves the activation of action-related representations, such that once activated, they influence response times (Glenberg and Kaschak 2002).

(3) Such that it inherits from the sensorimotor domain a structure that constrains the kinds of concepts we are capable of acquiring and shapes our conceptual schemes in a way that influences cognition (Gallese and Lakoff 2005; Lakoff and Johnson 1999; Lakoff 2014).

These approaches often rely on the same body of empirical material, which they cite as supporting their different interpretations. This chapter, as well as chapter 6, will be concerned largely with (3), which resists the notion that the sensorimotor system is modular, that concepts are modality-neutral, and that we comprehend them "on the basis of formal abstract models, totally unrelated to the life of the body, and of the brain regions governing the body's functioning in the world" (Gallese and Lakoff 2005, 455).

The type of scaffolding described in (3) exhibits a *nature-dependence* and *difference-making* relation between cognition and the body, such that the sensorimotor system gives rise to characteristic diachronic and synchronic influences on cognition that reflect aspects of the body. On the one hand, the specific morphological nature of our bodies and our sensorimotor system diachronically influences our possibilities for acquiring basic concepts and categories (Lakoff and Johnson 1999, 19; Lakoff 2014). These are not symbolic representations, and they are suitable only for beings with the type of body that humans possess (e.g., spatial ones like up, down, front, back, etc.). Moreover, in the course of the development of linguistic and cognitive skills, these concepts form the basis of an interconnected structure of metaphors that can influence cognitive activity. On the other hand, there are several synchronic influences because certain concepts are embodied. As Lakoff and Johnson (1999, 20) maintain, "An embodied concept is a neural structure that is actually part of, or makes use of, the sensorimotor system of our brains. Much of conceptual inference is, therefore, sensorimotor inference." Accordingly, when remembering, planning, and making decisions using abstract thought, cognitive activity is scaffolded onto a metaphorically organized conceptual structure that builds on parts of the sensorimotor system. The scaffolding structure, via metaphorically structured image schemas, exerts influence on cognition through the activation of source domains.

The next sections will offer more details on how complex scaffolding works, but before proceeding, it is important to draw attention to why such scaffoldings merit the *complex* label. Consider once again a rough-and-ready example analogous to the Botox example from the previous section. In contrast to Botox, when taking Tylenol to reduce physical pain (as fifty million Americans do each week), a different type of scaffolding is activated. Somewhat surprisingly, it turns out that Tylenol alleviates not only physical pain, but also social pain and anxiety, caused, for instance, by social isolation (DeWall et al. 2010). It is not hard to see that thinking in terms of simple scaffolding will not do justice to the Tylenol case. Instead, an explanation in terms of complex scaffolding is suitable. Most natural languages use pain expressions to convey the experience of social rejection ("You hurt my feelings!"; MacDonald and Leary 2005), indicating some kind of connection between physical pain and social pain. But the similarity between descriptions of physical and social pain is perhaps not merely contingent on linguistic conventions: the circuit for social pain and anxiety is scaffolded onto the system underlying physical pain (DeWall and Baumeister 2006; Way, Taylor, and Eisenberger 2009; Eisenberger,

Lieberman, and Williams 2003; Eisenberger 2012), such that undergoing one kind of pain can increase or decrease receptiveness to the other kind of pain (MacDonald, Kingsbury, and Shaw 2005) and that supportive social relations help decrease physical pain (Hoogendoorn et al. 2000). In such cases, beyond a mere simple scaffolding relation, there is a complex link between kinds of pain.

These examples epitomize the effects of active scaffoldings and illustrate two aspects that are central to this chapter: First, they indicate how an increased attention to scaffolding relations can help generate insight into the workings of the mind. Second, they suggest that different types of scaffolding and their different effects have potential implications for understanding and treating mental disorders. Nonetheless, unlike the relatively unsophisticated simple scaffolding, a full characterization of complex scaffolding requires giving an account of the processes of integration that explain the establishment of the relevant links. Without being able to offer an exhaustive catalogue, the following sections will depict two forms of integration that occur on different timescales.

3.2.2.1 Phylogenetic Integration In his reflections about his own work on the seemingly endless diversity of orchids, Charles R. Darwin (1862) commented on something that appears to be a general organizing principle in nature. He noted that when seeing a part of an organism as serving some function, we should not suppose that it was initially developed for this purpose. Instead, he emphasized that "the regular course of events seems to be, that a part which originally served for one purpose by slow changes becomes adapted for different purposes" (Darwin 1862, 346). Such cases of evolutionary integration of extant structures exemplify how features originally dedicated to a particular function can come to fulfill a different fitness-enhancing role (Gould and Vrba 1991). The role of such integration in the process of evolution is now widely recognized and thought to be occurring even at microscopic levels. *Integration* is a nonstandard term to apply in this context, but it is beneficial: it captures that an extant functional structure can be incorporated into and coordinated with a new functional structure, without ceasing to fulfill its original function.

When it comes to complex systems, integration takes on a specific form. In the course of evolution, relatively simple, extant structures often serve as the basis for more abstract ones (Dawkins 1976; Dennett 1995). If this is true in general, then it seems warranted to assume that complex and relatively recently evolved cognitive structures were built on structures that were initially evolved for other functions (Anderson 2010). We thus may expect that the machinery that drives human cognition can be accounted for by some combination of ancestral and derived structures. Indeed, neuroscientific research exploring different types of neural plasticity appears to support this thesis. Although it has long been assumed that brain function is best characterized in terms of local specialization ("anatomical modularity"), with parts of the brain dedicated to a single function (such as specific areas for language understanding in the left temporal lobe), it is now widely accepted that local specialization is less characteristic of cognitive function. Numerous regions assist several cognitive functions across domains, displaying the kind of plasticity that enables involvement in several networks that support diverse tasks.

To take an example that links the motor system to higher-order cognition, evidence from functional magnetic resonance imaging (MRI) and transcranial magnetic stimulation (TMS) experiments shows that a specific area of the left parietal lobe, which initially supports sensing fingers, at a later developmental stage sustains forms of mathematical cognition that require thinking in numbers (Penner-Wilger and Anderson 2013). The authors show that the neural structure responsible for sensing fingers has been "reused" for mathematical thinking and now supports both uses. This is achieved by demonstrating that areas of the brain associated with finger gnosis are activated while solving mathematical tasks and by showing that the disturbance of the relevant circuit impinges not just on finger gnosis, but also on cognitive performance. The upshot is that "the underlying relation between finger gnosis and math ability is indirectly causal, the result of a neural resource shared by finger representation and number representation, among other tasks" (Penner-Wilner and Anderson 2013, 1). This example illustrates how neural structures initially established for a particular purpose can be reused during evolution or normal development to support other functions while retaining their original functions. When possible, efficiency requires that existing structures are reused and integrated instead of the costly task of developing new ones. Different cognitive functions are supported by arranging the same neural circuits into different constellations. Compared to evolutionarily older functions, more recent functions tend to exploit a larger number of areas spread across the entire brain (Anderson 2010, 246–248).

In all, the fact that cognitive capacities conquered a neural niche in a structural platform largely tailored to control the body and online action (Anderson 2010, 2014; Dehaene 2009; Hurley 2008) may suggest further links and cross-domain effects between sensorimotor functions and cognition, opposing, for instance, a modality-neutral view of concept acquisition. Chapter 6 will offer a detailed investigation of such cross-domain effects and their potential implications for mental disorders.

3.2.2.2 Ontogenetic Integration Besides the phylogenetic processes that have shaped cognition, probably as a result of selection pressures (Williams, Huang, and Bargh 2009), the scaffolding of cognition onto the sensorimotor realm also occurs on a much shorter temporal scale. The acquisition, structuring, and meaning of certain concepts are scaffolded onto the human agent's sensorimotor capacities during typical development (Barsalou 1999, 2008; Glenberg et al. 2005; Niedenthal 2007). Some basic concepts through which human beings grasp their environment are directly "grounded" in specific characteristics of the body, while conceptual metaphors develop into cross-domain mappings (Lakoff and Johnson 1999; Lakoff 2014; Williams, Huang, and Bargh 2009). Basic concepts anchored in particular domains of bodily experience link bodily and abstract domains, while conceptual metaphors build on these concepts, further interlinking the two domains.

There is a straightforward way in which the physiognomy of our bodies influences concept and category acquisition (Lakoff and Johnson 1999, 33–35; Lakoff 2014). Image schemas like front/back and up/down are in a sense directly founded on the bodily experience of spatial location, and they provide a platform that multiple domains map onto. For instance, expressions like "prices rose" or "prices fell" exemplify how quantity judgments are conceptualized in terms of the sensorimotor experience of verticality ("more is up," "less is down"). For a more abstract example, consider how the domain of mind is comprehended via a broad-spectrum conceptual metaphor ("mental functioning is bodily functioning") that can be separated into subcases associated with separate source domains. For instance, the concept "understanding" is structured by the conceptual metaphor "understanding is grasping," while the concept "thinking" is structured by the conceptual metaphor "thinking is moving" (Lakoff 2014). These metaphors create strong associations with the bodily manipulation of objects, or with moving in space between an origin and a destination, and enable reflection that would differ if we "grasped" thinking in terms of something else, like "solving a riddle."

The metaphorical structure is also revealed in the way we portray character traits ("being an upright citizen"), as well as the way in which we depict our emotional lives. We describe feeling happiness as "being on top of things," implicitly conceiving the emotion at least partly in terms of simple vertical bodily orientations in space. In the same way, "feeling down" or "being in a deep depression" draws on the conceptualization "sad is down." Such metaphors are often constitutive of our understanding of emotions and integrate the bodily correlates of emotional experiences. For instance, anger expressed as "my blood is boiling" conveys the fact that bodily correlates of anger include elevated temperature and blood pressure (Kövecses 2000). Similarly, moral concepts arise from primary metaphors that connect experiences of well-being (feeling healthy, strength, upright position, purity, control, etc.) to moral vocabularies (Lakoff 2008, 2014).

Many complex metaphors arise from arrangements of more basic metaphors, some of which develop from regular correlations between embodied experiences in childhood. According to the developmental story, the young child goes through a period in which she fuses sensorimotor experiences with higher-order cognitive states, conflating intimacy and closeness or affection and warmth, and developing the relevant cross-domain associations (Johnson 1999). Via Hebbian learning—in which highly correlated activation patterns strengthen the synapses between the relevant neurons—the consistent coactivation of cognitive and sensorimotor domains leads to synaptic weight changes that establish permanent connections. For example, the metaphor "affection is warmth" arises due to children's experience of the stable correlation of affection and warmth when their caretakers habitually embrace them. The connections thus generated shape conceptual domains, and "neural mapping circuits linking the two domains ... constitute the metaphor" (Lakoff 2008, 26). The child at later stages of her development disconnects the different domains, but certain cross-domain associations endure, explaining other metaphors, like "a warm smile" or "a close friend" (Lakoff and Johnson 1999, 50; Lakoff 2014).

Consistent with the main thrust of this picture, but drawing on different sources, other researchers have suggested that basic concepts directly related to a child's physical experience of weight, height, closeness, and so on provide the platform for abstract concepts in a way that makes it possible for sensorimotor experiences to influence higher cognitive processes (Williams, Huang, and Bargh 2009). Although it is demanding for the child to process abstract concepts, more "concrete" ones linked to weight, height, closeness, and so on are relatively effortless because they are directly related to the experience of the environment. Admittedly, neither this theory nor the idea of a stage of conflation have been tested experimentally. However, the basic idea that children may acquire many metaphors when tracking stable patterns of correlation is appealing for several reasons. Given that these experiential correlations are based on universal aspects of the body and the world, the proposal potentially can explain how the same "metaphors have become enshrined in so many of the world's languages, without having to posit that the hundreds of ubiquitous mappings metaphor theorists have identified are all innate" (Casasanto 2014, 4; see also Casasanto 2013). That said, neither Johnson's theory of conflation nor Casasanto's position need to exclude the possibility that at least some mappings may be innate. This would be consistent with findings demonstrating some space-time mappings prior to linguistic experience (Srinivasan and Carey 2010; Lourenco and Longo 2011).

3.2.3 Extrasomatic Scaffolding

Although simple and complex scaffoldings are intrasomatic, some researchers stress that cognition depends heavily and in surprising ways not just on the body, but also on many different props, devices, and environmental structures to produce intelligent behavior. To elucidate further details of the framework and to introduce extrasomatic scaffolding, it is helpful to briefly revisit two accounts that offer "extended" and "enactive" views of cognition, which both have been stimulated by the acknowledgment that cognition can be "offloaded" in several ways to decrease the cognitive burden. Although offloading can occur in many ways, we may distinguish macro- and microlevel processes here.

As to the former, we may start by pointing out that evolved creatures actively engineer their habitats, constructing niches and structures (like dams or nests) that enduringly alter how the environment impacts them.

The environment thus reengineered affects the behavior of evolved creatures and increases their fitness, but also can transform them in quite substantial ways. An example for illustrative purposes is found in the development of human digestion (Sterelny 2010). To meet their nutritional requirements, great apes devote a significant amount of time to chewing food, supported by their strong jaws and teeth. In contrast, humans have started transforming raw nutrients by cooking, fermenting, chopping, slicing, grinding, pounding, and so on, which reduce not only the time spent chewing, but also the physiological demands of food intake on jaws, teeth, and guts. In a sense, the human digestive system has become environmentally scaffolded, enabling us to more fully exploit the nutritional resources of food at a lower physical cost. Some argue that the increasingly sophisticated preparation of food has not only changed our bodies (gut, skull, jaw size, etc.) but also assisted in the development of larger and more resourceful brains.

Further improving adaptive fit, human beings have constructed epistemic niches by reengineering informational structures of the environment that serve as scaffoldings for adaptive cognitive capacities and agency. Such environments, full of "props and other cognitive tools" that ease the informational demands of learning, allow the acquisition of substantial knowledge by means of apprentice learning and make available informational resources and learning across generations (Sterelny 2012, 36; 2010). Written language, numbers, calendars, and forms of notation are examples of resources that help acquire cognitive practices and the norms governing them in a process of "enculturation" (Menary 2015). The large number of shortcuts and engineered environmental features suggests that the key source of intelligent human behavior is not so much a relatively large brain size, but an adaptive habit "of offloading as much as possible of our cognitive tasks into the environment itself—extruding our minds (that is, our mental projects and activities) into the surrounding world, where a host of peripheral devices we construct can store, process, and re-represent our meanings, streamlining, enhancing, and protecting the processes of transformation that are our thinking" (Dennett 1996, 134–135). The habit of offloading liberates us from having to use costly internal solutions, and it produces scaffoldings for cognitive capacities that "squeeze maximum coherence and utility from fundamentally short-sighted, special-purpose, internally fragmented minds" (Clark 1997, 33; Wilson 2004).

Efficient intergenerational learning, the transmission of knowledge through an environment restructured to ease knowledge acquisition, requires the right kind of "apprentices." Suitable to epistemically engineered environments, and perhaps due to increasing climatic and ecological variability, human beings have evolved to prolong the developmental period accompanied by an adaptive developmental plasticity. But as neural structures are developmentally plastic, significant transformations in the environment do not leave these structures unaffected. Thus, to a certain extent, by continuously reengineering the epistemic environment, human beings have also reengineered themselves. Moreover, due to the cumulative nature of this continuous process, groups became more differentiated, with phenotypes reflecting both environmental dissimilarities and the learning history of their lineages. For example, most human beings have developed skills like hunting, but these skills tend to differ, reflecting local environmental differences (Sterelny 2003).

Cognitive processes involving offloading also occur on a microlevel. Frequently, using the informational structure of the environment, we pursue our goals by offloading and generating off-the-cuff scaffoldings on a much smaller scale. Adult human beings certainly are capable of systematically evaluating large quantities of information, applying logic and statistics. However, to reduce the cognitive load, adults often rely on cognitive shortcuts and heuristics, which are "strategies that ignore information to make decisions faster, more frugally, and/or more accurately than more complex methods" (Gigerenzer and Gaissmaier 2011). Heuristics save cognitive effort at the cost of accuracy but are able to do so by virtue of their being tailored to the configuration of the environment (Giegerenzer and Todd 1999). In some cases, shortcuts involve moving the body to decrease cognitive processing requirements. Consider a standard situation in a bookstore. You want to read the title of a book on the bookshelf, and you automatically tilt your head to get a better grip on its orientation. In such cases, the bodily movement is an adaptive alternative to the cognitively costlier mental rotation (Risko et al. 2014). Examples of such simple cognitive "offloading" onto the body abound, and the most illustrative ones involve memory tasks and mathematical and spatial reasoning (Goldin-Meadow 2003; Carlson et al. 2007; Chu and Kita 2011).

In many other cases, instead of moving our bodies, we offload onto the environment: we alter the environment such that the cognitive processing requirements of a task decrease. For instance, when recording information by writing it down or labeling objects in our homes, we are offloading and creating pointers that reduce the load on memory. The environment becomes a repository such that there is no need for an internal representation (Storm and Stone 2015; Tversky 2011). For another case of environmentally supported cognitive capacity, Sutton (2010) explores how small Elizabethan theatre companies dealt with having to perform many and rapidly changing repertoires. This meant that actors had to memorize around fifty new roles per year, and the question arises: How were they able to achieve this feat? The task was solved not by directly intervening in and extending the mnemonic capacities of the actors but by changing the environment to facilitate recall. Changes included altering the organization of the scenes, the interactions among actors, and even the thematic structures of plays.

More evidence comes from studies of home environments of elderly individuals, which offer easy access to information about the location of important objects and produce certain types of behavior. In fact, while *prospective memory* (remembering to implement planned intentions) significantly declines with age, the elderly actually perform well in their scaffolded environments (Schnitzspahn et al. 2011). In contrast, they become incapable of engaging in basic activities if moved to an unfamiliar environment, such as a hospital or nursing home.[8] It has long been known that due to difficulties with learning or memory, perception, and action control, elderly individuals increasingly tend to shift from internal to environmental control. Interestingly, cognitive difficulties increase when self-initiated, representation-heavy processes are necessary and decrease when the environment is organized in a fashion that makes salient the appropriate cues. These findings emphasize the need for more attention to the design of aging-friendly cognitive environments and assistive technologies (Lindenberger and Mayr 2014).

In a slightly more complicated process, we also offload self-control signal functions into the environment—for instance, by using knots for memory or tying string around our fingers. Such simple techniques of "intention offloading" produce environmental prompts for delayed intentions, assist in rising above capacity limitations, and fulfill otherwise unfeasible or very effort-intensive cognitive tasks. Surely this offloading results in increased sensorimotor efforts, but we tend to prioritize minimizing demands on cognition (Ballard et al. 1997). Opting for offloading instead of internal processing is influenced by our (mostly subconscious) evaluations of our internal and offloading capacities (Risko and Gilbert 2016; Nestojko, Finley, and Roediger 2013).

It is important to note that systematic and substantial offloading may lead to dynamic changes in the relevant cognitive capacity. Take, for example, our ability to offload and immediately retrieve information, which has dramatically increased with easy access to personal computers and the internet. The fact that we offload what we would have stored internally just decades ago has had consequences for memory skills. For instance, using the global positioning system (GPS) results in impoverished environmental encoding and lowered scene recognition and spatial memory such that it impairs wayfinding performance when the GPS aid is not available (Fenech, Drews, and Bakdash 2010). Another study compared individuals who merely observed an object with individuals who took a picture of the object. It was demonstrated that those taking a picture tend to unintentionally offload onto the camera and perform poorer on a memory test the next day (Henkel 2014).

3.2.3.1 Extended Cognition Cognitive niche construction thus includes offloading, modifying, and incorporating environmental structures, thereby augmenting cognitive skills. Some authors go further and maintain that offloading and much of our "online" cognitive activity is such that it undermines the idea that cognition only occurs within the confines of the head and the body of the individual. According to the *extended mind hypothesis*, cognition is not just causally affected by but is sometimes constituted by (often offloaded) bodily and environmental resources. Cognitive processes are thus sometimes ontologically distributed, such that external, nonbiological entities can figure as *constitutive parts* of cognitive processes (Clark and Chalmers 1998; Clark 2010a; Wilson 1994). Instead of merely occurring within the boundaries of our heads, forms of human cognition thus sometimes are constituted by feed-forward, feedback, and feed-around loops that cut across neural, bodily, and environmental resources (Clark 2008b, xxvi).

Although the main ideas have since motivated different accounts that can be described as *extended* (Sutton 2010; Menary 2007; Rowlands 2010;

Barrett 2011), the original version as defended by Clark rejects the focus on individuating cognitive states independently of features of the body and environment.[9] Some philosophers early on expressed doubts about whether the commitment to individualism allows for an account of content (Putnam 1975; Burge 1979). These accounts dealt with questions about whether the content of the states of two agents situated in two dissimilar environments can be identical, arguing that mental states may be internal but do not supervene on what is located within an individual's head. The extended cognition thesis is a more far-reaching form of externalism about the mind, sometimes described as *vehicle externalism*. It not only emphasizes the dynamic role of some environmental structures in accomplishing cognition, but also maintains that the vehicles of realization can individuate externally (see Hurley 1998, 2010). To explain what it means that cognition extends, Wilson and Clark draw on an illustrative parallel between cognitive extensions and house extensions that is worth quoting at length: Some of them

> are truly, massively, staggeringly transformative, while others are content to project a previously existing theme. Some are seamless and elegant, while others expose a barely connected hodgepodge of warring materials. Some are visibly permanent, built to persist as long as the main structure, while others are more temporary, apt for rapid dissolution or replacement. ... We should go further, however, and not simply think of cognitive and house extensions as alike but as instances of the very same kind of activity: the activity of modifying one's environment from the situation in which one find's oneself in ways that meliorate that situation. (Wilson and Clark 2009, 58)

In their seminal paper, Clark and Chalmers (1998) carve out their case by invoking two thought experiments that involve mental rotation and memory. In the second thought experiment, Inga consults her memory about the location of a museum, while Otto compensates for his deteriorating memory by checking the contents of the notebook that he always carries with him. The cognitively indistinguishable nature of these processes, Clark and Chalmers argue (1998, 14), supports the parity principle: if Otto's extended state (spread between Otto's notebook and his brain) takes up an identical functional role in Otto's mental life as Inga's internal state does in hers, then Otto's state is cognitive in the same way as Inga's state is (Horgan and Kriegel 2008). The parity principle brings out the commitment to *functionalism*, which defines mental states as multiply realizable functional states, specified in terms of relations between stimulus inputs, further functional states, and behavioral outputs. Building on such functionalist footing, extended cognition holds that if there is appropriate functional similarity between the external and internal cognitive elements, then both should be granted equivalent cognitive status (Wheeler 2010). To delineate the domain of the cognitive, Clark uses the folk-psychological knowledge of coarse-grained functional roles of cognitive processes that guide intelligent behavior. The folk notion is locationally uncommitted and "is only counterintuitive ... if we are already in the grip of a form of theoretically loaded neurocentrism" (Clark 2008b, 105).

Overall, extended cognition can be seen as an attempt to reconcile computationalist with dynamicist approaches (Clark 2008a, 55), so it is in some sense appropriate that Wilson (1994) has referred to the main idea as "wide computationalism." The computational system may sometimes incorporate parts of an organism's body and environment. Moreover, the extended mind hypothesis also emphasizes that several online cognitive processes do not require internal representations, without denying that other cognitive processes are representational. One main difference from cognitivism is that the origin of representations and cognitive states is not located in the brain, but "may supervene on activities and encodings that crisscross brain, body, and world" (Clark 2008b, 149).

3.2.3.2 Enactivism Enactivism is often associated with the sensorimotor approach to visual perception (see, e.g., Noë 2004; O'Regan and Noë 2001), which holds that sensorimotor capacities for perceptual experience and for intentional agency are essentially interdependent (Ward 2016). In this context, *enactivism* refers to an integrative theoretical framework for cognition, sometimes described as *radical embodiment* (Thompson and Varela 2001). In their pivotal work, Varela, Thompson, and Rosch (1991) present enactivism as an approach that understands perception and cognition as oriented toward action, emerging from and composed of configurations of dynamic interaction between organism and environment. Since then, the view has been developed in various ways (Gallagher and Varela 2003; Chemero 2009; Di Paolo, Rohde, and De Jaegher 2010; Gallagher 2017; Hutto and Myin 2013; Thompson 2007; Ward 2016).

Proponents of extended cognition and enactivism agree on numerous points—for instance, that instead of decomposing the cognitive system into inner modular subsystems, cognitive science should identify and investigate dynamical systems that transgress the traditional divisions between brain, body, and world. For example, Varela, Thompson, and Rosch (1991) maintain that some of the robots from Brooks's research (mentioned in chapter 2) are structurally coupled to, and thus inseparable from, the environment. Relatively simple robots with a limited number of independent sensors that measure the distance to the nearest object successfully navigate a cluttered environment. Such robots are coupled with a world that contains only very specific stimuli that they react to, forming a closed system in which only particular stimuli provoke reaction, which, in turn, determines the significance of those stimuli for the robot (for a discussion, see Chemero 2009). The interaction with relevant environmental offerings replaces the need for representational processes. To Varela, Rosch, and Thompson (1991, 212), such examples help show that enactivism is not merely a philosophical thesis, "but the result of forces internal to research in cognitive science, even in the case of those hard-nosed engineers who desire to build truly intelligent and useful machines."

Although extended cognition offers ideas and tools that can be used in conjunction with more traditional approaches to studying cognition, enactivism holds a more radical thesis and neither commits to functionalism nor aims at reconciliation with computationalist approaches. Instead of an abstract conception of cognition, enactivists start out with the observation that organisms are self-organizing and autonomous systems. They maintain that "life and mind share a set of basic organizational properties, and the organizational properties distinctive of mind are an enriched version of those fundamental to life" (Thompson 2007, 128; Di Paolo, Rohde, and De Jaegher 2010). Enactivists locate life and mind on a continuum and model cognition on biological, and not computational, processes. The aim is to provide an account of cognition that can link principles at different levels of complexity and account for organizational properties and behavioral patterns in simple life-forms.

Enactivism maintains that explaining much ordinary cognition does not necessitate using the notions of representation and computation. In fact, these notions misleadingly describe cognitive processes as more intellectualistic than they really are. An often-used example to illustrate the firm coupling between an agent and particular environmental features is running to catch a flying ball. According to a broad cognitivist view, the visual system conveys information about the current position of the ball to the cognitive system, which then calculates the trajectory of the ball. In contrast, enactivists emphasize that we use a less demanding solution: we couple perception to an invariant in the optic flow and adjust our running pattern so that the ball's flight always appears straight compared to the background. Thus, instead of acquiring exact representations that allow specific internal calculations, agents create an agent-world circuit that is particular to this task and that enables environmental features to continuously guide running (McBeath, Shaffer, and Kaiser 1995; Clark 2007).

Critics of this view have pointed out that though enactivism explains basic forms of cognition (such as basic coordination and categorical perception), it fails to scale up to more complex cognitive activities that characterize human beings and that presumably require representations. These tasks, so the criticism goes, often involve thinking about past, future, and nonpresent objects, events, and persons, which simply cannot be accomplished without representations (Clark and Toribio 1994). One reply to this objection is to show that enactivism has the resources to explain tasks that are widely thought to heavily involve representation (see van Rooij, Bongers, and Haselager 2002; for a discussion, see Chemero 2009, 40). Others note that for enactivism "to scale up to cognitive states such as imagining or remembering, the brain doesn't decouple or recreate a process that was representational to begin with; rather, the system (using the same motor control or forward control mechanism) enacts (or re-enacts) a process that is now coupled to a new cognitive action. In remembering, for example, there may be reactivation of perceptual neural processes that had been activated during the original experience" (Gallagher 2017, 15).

Another point of divergence is that enactivists regard cognition not primarily as extended, but as "extensive" (Hutto and Myin 2013, 7). The point is that the notion of "extensive" is free of some of the problematic implications of "extended"—that some cognitive system can be enlarged into the environment. For a fuller explanation of cognitive skills, one should focus on the history of previous engagements and interactions of the organism, without starting from an internalist picture of the mind, which then can be extended (Hutto and Myin 2013, 47). That is, "fundamental cases of perceiving and thinking are not necessarily grounded in, nor do they take the form of, representing aspects of the world or having contentful thoughts about those aspects. And to think this aligns perfectly with understanding cognitive processes as a form of wide reaching activity that is—at root—extensive and unbounded; thus extensive minds are not merely, occasionally and in special circumstances, extended" (Hutto, Kirchhoff and Myin 2014, 2). Reaching beyond extended cognition, Hutto and Myin (2013) critically assess arguments in favor of internalism about phenomenality and argue that phenomenality might also be extensive. In contrast, proponents of extended cognition maintain that the minimal supervenience base of phenomenal properties is restricted to the brain. Both Clark (2009) and Chalmers (2008, x–xi) have expressed strong reservations about the thesis that conscious mental states extend into the environment. Hutto and Myin do not deny the claim about the minimal supervenience base, but they maintain that it is not warranted to infer that a full understanding of phenomenal experience could be attained by studying only neural properties (Hutto and Myin 2013, 164). Also, in contrast to extended cognition (functionalism), radical enactivists claim that bodily processes shape and contribute to the constitution of consciousness in an irreducible and irreplaceable way.

3.2.4 Intersomatic Scaffolding

Now that we have offered a brief discussion of offloaded extrasomatic scaffoldings, as well as extended and enactive views that place crucial weight on these structures, we may further elucidate the third form of scaffolding within ASC's framework, *intersomatic scaffolding*. Both extended cognition and enactivism claim that, when appropriately coupled with active features of the environment, cognition can be, with or without conscious awareness, distributed across brain, body, and world in an ontological sense. But perhaps due to the way in which technological inventions have changed our lives within the last couple of decades, large parts of the literature tend to assume that the most obvious contenders for cognitive extension are physical entities in the immediate environment that increase problem-solving abilities (see Clark 1997, 32–33). The most well-known elaborations of how cognitive extension works largely involve objects and devices like notebooks or cellular phones, although influential authors accept not only that cognitive tools (including specific mathematical and social schemata) are passed down to us in language and culture, but also that the cognitive capacities of other people may form social scaffolds for cognition:

> We keep "pointers" and "indices" in our brains and leave as much of the actual data as we can in the external world, in our address books, libraries, notebooks, computers—and, indeed, in our circle of friends and associates. (Dennett 1996, 143–144)

> What about socially extended cognition? Could my mental states be partly constituted by the states of other thinkers? We see no reason why not, in principle. In an unusually interdependent couple, it is entirely possible that one partner's beliefs will play the same sort of role for the other as a notebook plays for Otto. (Clark and Chalmers 1998)

Thus sophisticated cognitive skills can be technologically, but also socially, scaffolded (see also Chalmers 2008; Wilson and Clark 2009). Wilson (2005, 230) maintains that cognition may sometimes extend and highlights "the constitutive role that an individual's social milieu plays in her cognitive activity." Nonetheless, there are forms of intersomatic, social scaffolding that are important for understanding phenomena in psychopathology but are largely neglected in the literature and that cannot be accommodated readily by enactivism or the extended cognition framework. There are two key issues to note here.

First, though some of the original extended cognition proponents support the idea that cognition can be socially extended, researchers in the field since have been more focused on forms of nonsocial support to examine ways in which cognition at times can extend. Most discussions use physical props, as mentioned previously, which underplays how often problem solving is attained through social interaction. This is surely due to a combination of several factors, but the underlying assumption is that cognitive extension, whether physical or social, *adheres to the same regularities* and can be explained within the same framework of extended cognition. However, although they can be difficult to separate in complex environments, we shall see that social scaffolding and physical scaffolding can be significantly distinct. The parity principle provides a helpful test to escape brain-centered prejudice about the location of the vehicles of cognition, but also makes it easier to oversee how social scaffolding can transform cognitive processes in unique ways.

Second, a few researchers have explored the question of socially extended cognition in different ways, typically investigating larger problem-solving cognitive systems and exploring, for instance, cognition in larger groups (Theiner, Allen, and Goldstone 2010; Tollefsen 2006; Sutton et al. 2010;

Gallagher 2013a). Some of the inspiration for these accounts comes from seminal work by Edwin Hutchins (1995a, 1995b, 2011, 2014), who has investigated cognitive systems that encompass multiple agents and artefacts. Hutchins pointed out that such systems have unique cognitive properties. For instance, in the case of navigating a large ship, the crew itself—together with certain nonmental instruments like sextants, alidades, and charts—builds a distributed cognitive and computational system (Hutchins 1995a, xiv). In this scenario, cognition in individuals constitutes parts of the larger cognitive system. The tasks are distributed such that the instruments allow the crew to perform actions that they excel at, such as "recognizing patterns, modeling simple dynamics of the world, and manipulating objects in the environment" (Hutchins 1995a, 155). Others have argued that groups of people can function as distributed, "transactive" memory systems, such as couples who display particularly integrated cognitive processes (Sutton 2010; Huebner 2014; Harris et al. 2014). Wegner, Erber, and Raymond (1991; see also Wegner 1995) maintain that members of such social constellations develop mostly tacit suppositions about the functional profiles of group members and distribute the cognitive burden by assigning different parts of the overall information processing load to each other.[10]

The kind of intersomatic scaffolding that this section aims to draw attention to exhibits important differences from cognitive systems made up of human-artefact cognitive systems on the one hand and human-human cognitive systems on the other. *Intersomatic scaffolding* refers to a form of socially extended cognition, but aims to describe cases that require a departure from some of the assumptions in the extended cognition framework. The cognitive processes that intersomatic scaffolding targets are typically second-person, face-to-face, bodily interactions that have a key role in a number of psychopathological phenomena. A good example that chapter 6 explores in detail is interpersonal (dyadic) emotion regulation, in which the cognitive scaffolding not only includes the agent's sensorimotor apparatus, but also extends into the environment to include parts of the interacting body of another human being. Such processes, as will be argued in detail in chapter 6, do not adhere to the same regularities that standard extended cognition cases involve. Also, they are very different from cognition in larger groups: they do not require the interactors to possess a common goal, to develop knowledge about the areas of expertise of others, or to distribute the cognitive burden accordingly.

Intersomatic scaffolding is a form of socially extended cognition that cannot be readily accommodated in the extended cognition framework. In some ways, enactivism would be a better fit, but not its commitment to the idea that mind and life have in common fundamental organizational properties. First, intersomatic scaffolding involves no commitment to functionalism, which makes it unproblematic to maintain that these types of interactions fulfil a function that is realizable by other means, without necessitating the physical body of another individual. Second, it is less committed to representations and emphasizes that these cases of socially extended cognition exhibit irreducibly emergent (and sometimes diachronically novel) cognitive properties that the standard account of extended cognition is unable to accommodate. These properties are systemic, instantiated on the system level, without following from the behavior of its parts (for a more detailed discussion, see Varga 2016).

3.3 Conclusion

Although the turn toward approaches that emphasize embodiment may not constitute a genuine paradigm shift, it is nevertheless a research program that has exerted significant influence on the course of cognitive science. This chapter introduced and motivated a framework that integrates a medley of positions traditionally subsumed under the label of EC, along with other accounts that stress forms of nontrivial dependence of cognition on the body and the environment. It not only opposes that cognition is amodal and occurs entirely separated from the sensorimotor system, but also maintains that cognition and language use are *actively* scaffolded onto the environment and sensorimotor processes. ASC incorporates extant accounts that endorse at least one of the two claims of inseparability, granting explanatory relevance to the body and its engagement with the environment. The aim of the taxonomy of forms of scaffolding was not to offer a comprehensive ontology, but to provide distinctions that could be productively applied to the context of psychopathology.

This goal will be pursued in chapters 6–8, but, prior to that, chapter 4 aims to locate ASC in the landscape of current discussions in the philosophy of cognitive science. This will involve addressing not just epistemological but also ontological separability, an issue that we have set aside. Although the focus of this book is pragmatic, focusing on how forms of active scaffolding can be exploited to understand and potentially alleviate disturbances associated with various mental disorders, some of the debates about the extent to which mind and world are intertwined cannot be avoided. Engaging in some of these debates will also help assess the viability of a type of *pluralism,* motivated by a suspicion that no single account of cognition is both broad and specific enough to account for the wide variety of cognitive processes.

4 Actively Scaffolded Cognition, Ontological Inseparability, and a Pluralist Landscape

The taxonomy offered in the previous chapter raises several questions about the so far relatively vague relationship of nature-dependence that holds between the scaffold and what is scaffolded. Although the forms of scaffolding under the ASC umbrella all support the epistemological inseparability thesis, the question concerning inseparability in the ontological sense now forces itself upon our attention. Consider simple scaffolding, for instance. Those leaning toward more radical accounts (e.g., extended cognition, enactivism) will likely insist that the muscle activity in the gebellar region is a part or component of the machinery that underlies the relevant emotional state. Those with less bold leanings would contend positing a simple causal relation. In the case of complex scaffolding, those leaning toward more radical accounts would perhaps maintain that the sensory-motor circuits (re)used for cognitive activity are a part of the cognitive activity and not mere background conditions. Presumably, the status of intrasomatic scaffolding will be disputed along similar lines.

Assuming that causation and constitution are distinct metaphysical relations, it cannot be the case that both sides are right. Moreover, positioning ASC with respect to this issue gains additional importance because predictions of how the relata will behave in causal versus constitutive relationships of dependence differ. This chapter and the next will therefore shoulder the task of exploring this question while also providing further specifications of other aspects of ASC. The enterprise aimed at meeting this expectation might take two paths. The first option, investigated in the remainder of this chapter, is to provide an extensional characterization of ASC along the lines of the debate on boundaries of cognition. The second option, explored in chapter 5, is to explore questions about the status and boundaries of ASC in light of more general questions on how to delineate the boundaries of mainly biological systems and mechanisms.

4.1 Boundary Disputes

EC has reached a particular stage of maturation. The main idea that cognition is embodied in some form does not require much additional defense beyond explanatory success, and the main arguments in various accounts do not focus primarily on the criticism of cognitivism but on further elaborating the basic principles that explain the embodied nature of cognition.[1] Nonetheless, the debate between cognitivism and EC, as well as between EC accounts, continues, and it often revolves around the question of the boundaries (the "where") of cognition, which is intertwined with the question of ontological inseparability.

Numerous philosophers maintain that the debate requires providing a "mark of the cognitive" that picks out all and exclusively cognitive processes (Adams and Aizawa 2010; Rowlands 2008, 2010; Adams and Garrison 2013; Walter and Kästner 2012).[2] One idea is that comparing the functional profiles of organism-external parts or processes with internal processes and maintaining that organism-external process X is the *same* process as intracranial process Y involves at least implicitly taking a stance on what kind of differences between the two processes we can abstract from. Others are skeptical about the need to provide such a mark. Clark (2011, 4) maintains that the question about parity is one that "we can perfectly well ask ourselves without first adopting any explicit 'mark of the cognitive.'" When investigating problem solving, memory, or spatial reasoning, a mark of the cognitive is not necessary, because there is widespread agreement that these processes count as cognitive (Shapiro 2011, 186). But though this may be true in some standard cases, the picture is less clear when it comes to assessing the cognitive status of emotions, forms of instinctive reactions, comprehension, and consciousness (Walter and Kästner 2012; van Holland 2013). Moreover, one could accept this point but still maintain that detailing what might underlie such an agreement could be productive.

The first part of this chapter will engage two accounts of the mark of the cognitive, by Mark Rowlands (2010) and Adams and Garrison (2013). Rowlands (2010) provides sufficient conditions for something to count as cognitive, whereas Adams and Garrison (2013) offer a necessary condition. Both accounts hold that mere information processing does not constitute cognition, and neither aims to *naturalize* the mind in the sense of offering a reductive definition of cognitive processes in noncognitive terms. Whereas the engagement with these accounts will unearth some decisive challenges that attempts at demarcating cognition should address, the second part of the chapter will explore what we could call the *fine-grained vs. coarse-grained debate* (Sprevak 2009).[3] Those leaning more toward cognitivism and weaker accounts of EC argue in favor of a *fine-grained view*, maintaining that the boundary of the cognitive should be fixed by examining the fine-grained functional details of cognition that much conventional cognitive psychology investigates.[4] In contrast, those leaning toward stronger forms of EC typically favor a *coarse-grained view*, which abstracts from fine-grained characteristics and focuses on larger cognitive ensembles, often justified by maintaining that this level of analysis is both consistent with commonsense psychology and avoids conceptualizing cognition in an overly anthropocentric manner.

The debate has come to a halt. Clark thinks that the coarse-grained approach is clearly intuitive and consistent with common sense, but critics have opposing intuitions, thinking that the coarse-grain parameter and the location of cognitive activity are at odds with common sense. To understand how this impasse ensued and to explore how progress could be made, this chapter will investigate the source and possible implications of the clash of intuitions, while bearing in mind that the failure of rigorously trained professional philosophers to elicit converging intuitions about a circumscribed subject might indicate a deeper problem. To shed light on this issue, the chapter will examine the nature of the relevant philosophically significant concepts, such as "cognition" and "mind." It will be argued that the dispute does not warrant eliminating the concept "cognition" from scientific taxonomy, but can be regarded as an opportunity to explore the possibility of a kind of pluralism that is familiar from mature sciences and that is a good fit for ASC. The closing part of the chapter will respond to two potential concerns, according to which pluralism implies antirealism and impedes scientific progress.

4.2 Two Attempts at Demarcating Cognition

Adams and Garrison (2013) draw on the intuitive idea that the behavior of creatures may involve some information processing without qualifying as properly cognitive. Consider Rodney Brook's robot, Herbert. The robot is able to find and collect empty soda cans and process information about the environment, much like bacteria that are able to move away from toxins, or slime mold that can find the shortest way to food through a maze. However, according to Adams and Garrison, the processing at stake is noncognitive, because the information processers do not act for "reasons" of their own, and their behavior is not based on the mechanisms of beliefs, desires, and intentions. Herbert can perhaps be said to have "reasons" that derive from the designers, but in all these cases, there are noncognitive mechanisms that explain the relevant behavior in nonrepresentational, biological, or computer-scientific terms.

In contrast, the behavior of creatures that cognitively process information act for reasons such that explaining their behavior requires recourse to reasons. As Adams and Garrison (2013, 346–347) maintain, cognitive status and reason explanations require that the goal and the strategy for accomplishing it by certain activities be represented within the system. Unlike the cases of bacteria, mold, and Herbert, when a cat moves toward a bird with the goal of catching it, the cat has recourse to a representation or model of the goal and the means to achieve it. Adams and Garrison maintain that the desire for food and the belief that the bird is edible are, in this case, the mechanisms that produce the relevant behavior. These reasoning mechanisms might have been selected for, but they nevertheless constitute individual reasons for the cat to catch the bird, emanating from its beliefs and desires. In other words, to reach the level of cognitive information processing, representational states must be contentful: it must be possible for them to be falsely tokened and to misrepresent the actual states of affairs (Adams and Garrison 2013, 350). In brief, for Adams and Garrison, the activity of a system S does not qualify as properly cognitive if S's behavior is not amenable to a "system-centered" reasons-based explanation. Conversely, if explaining S's behavior involves rising to the level of reason explanation, then S is properly cognitive.[5]

The account put forward by Rowlands operates with a notion of cognition that includes perceptual processing and offers sufficient conditions for a process to qualify as cognitive. According to Rowlands (2010, 110–111), a process P is a cognitive process if

(1) P involves information processing—the manipulation and transformation of information-bearing structures;

(2) this information processing has the proper function of making available either to the subject or to subsequent processing operations information that was, prior to this processing, unavailable;

(3) this information is made available by way of the production, in the subject of P, of a representational state; and

(4) P is a process that belongs to the subject of that representational state.

Discussing the Gibsonian account of visual perception, Rowlands argues that condition (1) is met. On a common account of "information" perception (X conveys information about Y only if the probability of X given Y is 1, or X conveys information about Y only if there is at least a significant increase in conditional probability) involves information processing. Moreover, using Millikan's etiological concept of proper function (Millikan 1984, 2004), Rowlands argues that the proper function of visual information processing is to make information available for further processing.[6] To meet condition (3), Rowlands adopts an inclusive notion of representations that only requires adequacy conditions.

While conditions (1) to (3) are relatively uncontested on most accounts, condition (4) seems more controversial. To avoid the threatening circularity, Rowlands (2010, 135) adds that a subject in the sense required by condition (4) is an individual that owns processes that satisfy conditions (1)–(3). The real task thus is to elucidate how a subject can genuinely "own" its subpersonal and personal-level cognitive processes. Because the ownership of subpersonal processes derives from the ownership of personal cognitive processes, the task is to spell out the conditions of ownership for personal-level processes. In short, a reliable indicator of a subject S's ownership of a personal-level process P is that S has epistemic authority over P.[7] Although one does not enjoy such authority over the processes that we lack epistemic access to (e.g., the processes that convert raw primal sketches into full primal sketches), they can qualify as cognitive "because of the contribution they make to—the ways in which they are integrated into—personal-level cognitive processes to which the authority criterion is applicable" (Rowlands 2010, 154).

Rowlands continues by arguing that the basis for ownership of cognitive processes is provided by the idea that basic coping and higher-order cognition are both forms of disclosing activity and that disclosure is always disclosure to a subject. In other words, cognitive processes are essentially owned because revealing activity is essentially owned. This then allows

Rowlands to argue that cognitive processes can be extended because they belong to the class of revealing activities, which do not necessarily stop at the organism's head or skin. Because the account risks stretching the notion of disclosure too thin, Rowlands points out that not all disclosing activity is cognitive. Instead, cognitive disclosure is disclosing activity that satisfies the conditions (Rowlands 2010, 212).

4.2.1 Test Case: Emotion Regulation

Although these accounts display notable differences in the type of criteria they offer, both hold that mere information processing does not constitute cognition. In ensuing discussions, some remain unconvinced that Rowlands's account is shielded against running into a vicious circle (Adams and Garrison 2013), whereas others argue that the availability of a reasons-based explanation for behavior is not a necessary condition (Elpidorou 2014). Both accounts capture highly characteristic features, but it is helpful to at least point to types of cases in which they would generate incorrect conclusions. For this, the following sections will reach beyond the typical examples involving artifacts and consider cognitive processes that involve social interaction. Any account aiming to specify what constitutes cognition (and perhaps also the conditions it requires to occur) needs to accommodate the kind of characteristics that such processes display.

Consider emotion regulation in infants. Whereas adults have access to several conscious emotion-regulation strategies (cognitive reappraisal, response modulation, etc.), emotion regulation in infants, who have not yet mastered these cognitively mediated strategies, can be self-oriented (e.g., thumb-sucking), or occur in a complex interaction with the caretaker. In the latter case, the infant and the caretaker are understood as components of a dynamically interacting dyadic system (Sameroff 2009; Leclère et al. 2014; Feldman 2007; Cole, Hall, and Radzioch 2009). Importantly, the infant's emotional states are dyadically regulated: emotion regulation is accomplished within the framework of the dyadic interaction, but synchronized patterns of bodily interaction serve as the vehicle of the regulation. The interaction with the caretaker figures as a genuine component of the infant's emotion regulation (Tronick 1998, 296). Chapter 6 will say more about why we should think of this as a case of extrasomatically scaffolded cognition; in the context of this chapter, it suffices to assume that this is correct and focus on exploring whether the conditions proposed by Adams and Garrison and Rowlands are met.

As the infant distinctively seeks interaction and continuous engagement with the caretaker, a reasons-based explanation could maintain that the *reason* infants engage in such tight bodily interaction is to attain regulation. However, this appears to be the wrong kind of reasons-based explanation: it neither cites the infant's "own" reasons nor requires that the infant represents the goal of achieving emotion regulation and the strategy for accomplishing it. In fact, this explanation would be comparable to the one formulated in the following sentence: "The reason plants turn their leaves toward the light is to maximize photosynthesis." Invoking the infant's purported desire (to be near the caregiver) and belief (about the "strategy" to apply to attract the caregiver's attention) would cite states that are not directly relevant to the infant's emotion regulation, while appealing to the infant's representation of the goal (to achieve emotion regulation) together with the representation of a specific strategy (bodily interaction) to accomplish it would require attributing a cognitive sophistication that the infant does not yet possess. Consequently, emotion regulation thus achieved is not amenable to an explanation that invokes the infant's reasons, states, and representations.

The bodily interaction that functions as the vehicle of emotion regulation fails to exhibit the kind of agency and epistemic authority that for Rowlands characterizes personal-level cognitive process, but it may still qualify as cognitive. This would require showing that (a) the infant's emotion regulation constitutes a properly owned person-level cognitive process and (b) the bodily interaction constitutes a properly integrated subpersonal process. To (a), if epistemic authority constitutes a criterion for our ownership of person-level processes, and if "disclosure supplies the ultimate basis for our ownership of cognitive processes" (Rowlands 2010, 163), then the infant's emotion regulation may qualify as cognitive because it may be understood as a disclosing activity. After all, changing the duration or intensity of an emotional state tends to alter the way the world appears to us.

To (b), the bodily interaction in question may constitute a properly integrated subpersonal process if it is akin to saccadic eye movements, which are disclosing activities that supply information to personal-level processes. But does the bodily interaction contribute to personal-level processes by making information available? Perhaps the bodily interaction enables transmission of information about the caretaker's body, emotional state, or perhaps proprioceptive information. But it is not clear why the presumed representational states carry *task-relevant information* that enables emotion regulation in virtue of this information. Although the bodily interaction may qualify as a subpersonal process that contributes to further processing, it is not the kind of disclosing activity that meets Rowlands's conditions.

4.3 Coarse- versus Fine-Grained Analysis

Another way to think about what makes a process cognitive is to consider the functional role it plays in an organism's mental life. Proponents of extended cognition argue that such analysis must adopt a coarse-grained functional approach, which, they maintain, would support the EMH. For two organisms O_1 and O_2, if state X interacts with O_1's desires and beliefs in a similar way to the way in which state Y interacts with interacts with O_2's desires and beliefs, then both processes should have an equal claim to be cognitive, even if X's physical supervenience base lies within O_1's body, while Y's physical supervenience base includes entities or processes outside of O_2's body. Although this requires abstracting from fine-grained characteristics, critics argue that the boundary of what is considered cognitive should be fixed by examining fine-grained details that much conventional cognitive psychology investigates (Adams and Aizawa 2008, 2010; Rupert 2004, 2010).[8] Inga's and Otto's cognitive processes are simply too different to belong to the same kind: not only does Otto's case involve motor-processing systems and require his visual system and additional beliefs about the notebook, but his "memory" also lacks characteristic fine-grained features of human memory (generation effect, Rescorla-Wagner law, etc.).[9] For instance, it lacks "negative transfer," a widely exhibited form of interference that appears when past learning affects a subject's capacity to learn and remember new associations.[10]

Supporters of the EMH acknowledge the existence of fine-grained functional differences but resist the conclusion that the failure of "extended" processes to exhibit such fine-grained characteristics warrants denying cognitive status. Along similar lines, Wheeler (2010) argues that cognitive scientific research does not require cognitive processes to be individuated on a fine-grained level. Wheeler asks us to imagine a human being who exhibits context-sensitive storage and retrieval of information but whose memory system does not display the generation effect. His point is that "neither commonsense nor cognitive psychology baulks at the thought that this subject's feats should count as genuine cases of remembering" (Wheeler 2010, 257). Thus, exhibiting the fine-grained feature cannot count as a defining, but rather an accidental, feature of human memory: "The difference between exhibiting or failing to exhibit fine-grained functional traits (like the generation effect) doesn't mark the boundary between being a cognizer and not being one" (Wheeler 2010, 257).[11]

Whereas Wheeler points to scientific psychology as the appropriate basis for circumscribing the functional roles of mental states, Clark emphasizes the folk conceptions of these roles.[12] Clark maintains that our implicit folk psychology is coarse-grained, not merely because the fine-grained aspects are established by psychological research unknown to the "folk," but also because it is "designed to be insensitive to any differences in states in the head that do not issue in differences in quite coarse-grained behaviour. ... It does so because its purpose is to provide a general framework in which gross patterns of the behaviour of many other well-adapted beings may be identified and exploited" (Clark 1989, 48). Moreover, folk psychology is coarse-grained to the extent that it is compatible with EMH claims about location and vehicles of cognition (Clark 2011, 451; 2008b, 105). With this in mind, let me draw attention to two central passages that are worth quoting at length:

> Should we fix the domain of the cognitive by reference to the actual (detailed) processing profiles of normal human agents (deferring, I suppose, to our best final science of the normal human brain)? Or should we count ourselves as already commanding an understanding capable of extension to new cases? The argument by Clark and Chalmers assumes that we do possess some such understanding, and that it is rooted, roughly speaking, in our implicit knowledge of the distinctive functional role of cognitive processes in guiding intelligent behavior. It is this knowledge that allows us to count alien processes in nonhuman animals as properly cognitive, and on which we must rely when applying the informal test embodied in the parity principle. The alternative (making everything depend on identity with processing in the normal human case) strikes us as both anthropocentric and ultimately unworkable. (Clark 2010a, 54–55)

> What makes a process cognitive, it seems to me, is that it supports intelligent behavior. ... The notion of a cognitive process, if that is correct, is best unpacked as the notion of a process that supports certain kinds of behavior. Why ask for more? (Clark 2010b, 93)

These key passages shed some light on two crucial claims. The first claim is that folk psychological knowledge is coarse-grained, "rooted" in some implicit knowledge of characteristic functional roles that different cognitive processes play in guiding behavior. Knowledge about the realm of the cognitive that guides our ability to identify certain processes as cognitive stems from our implicit knowledge of a certain repertoire of functional profiles. If a process P fits this profile, then P is a cognitive process. However, avoiding an unrestrained proliferation of cognitive processes, if P fails to fit this profile, then P is not a cognitive process. If I understand this correctly, the claim can then be formulated along these lines:

(1) *Ascription of cognition based on functional profiles:* A process P qualifies as cognitive only if P occupies a distinctive functional role (linking sensory input and intelligent behavioral output) that matches our coarse-grained folk psychological repertoire.

The second claim starts from the observation that our folk psychological knowledge is coarse-grained, but it has the goal of establishing how the grain parameter should be set. Clark invites us to imagine an encounter with intelligent Martians. Somewhat echoing earlier work by Lewis (1980) and Block (1991),[13] Clark argues that it would be "anthropocentric" and counterintuitive to claim that intelligent Martians with a dissimilar biological makeup could not have genuine cognitive processes. Clark takes this to show that the vehicles of realization for specific functions are irrelevant for determining cognitive status, so long as their physical material is organized in a manner that results in the appropriate functional profiles. In part because the Martian experiment is deemed plausible on independent grounds, Clark suggests that we should let the "Martian test" determine the appropriate grain level to test functional similarities. In other words, the claim is as follows:

(2) *Martian-friendly grain parameter:* We should adopt a functionalist account with a grain parameter set sufficiently coarse to accommodate the Martian intuition and thus allow intelligent Martians to qualify as cognizers.

In a final step, not under consideration here, Clark argues that a grain parameter coarse enough to allow for intelligent Martians to have cognition will in turn support the EMH: if the Martian's processes count as cognitive, then there is no principled reason that Otto's extended process should be denied cognitive status. As Clark (2010b, 89) puts it, "If we accept the Martian memory into the cognitive fold, surely only skin-and-skull based prejudice stops us extending the same courtesy to Otto" (see also Sprevak 2009). Although Clark's overall account contains much more detail than I can do justice to here, the focus was on identifying two central claims for further elaboration, the success of which are crucial for the success of Clark's overall argumentative strategy. The following sections will investigate to what extent claims (1) and (2) are supported. Some challenges to claim (1) will be noted, but the focus will be on claim (2).

4.3.1 Ascription of Cognition Based on Functional Profiles

To take a closer look at claim (1), let us consider the Marian scenario. Imagine that we encounter Clark's Martians and attempt to establish whether they are cognizers like us. Following Clark, our approach is *process-first*, starting with deciding whether the process, P, that drives the behavior of the Martian subject (MS) constitutes an instance of cognitive processing (as opposed to noncognitive processing). Focusing on the role of P in guiding MS's behavior and disregarding facts about how P is realized, we draw on our implicit knowledge of functional profiles and judge that P is cognitive to the extent that P connects MS's sensory input and behavioral output in a way that fits our folk-psychological repertoire. Consequently, if P fails to fit our repertoire of profiles, our first conclusion is that P is not cognitive. Moreover, if our prolonged observation reveals that none of MS's processes fit our repertoire of functional profiles, we conclude that MS is not a bearer of cognitive processes.

Nonetheless, at least in some cases, our approach appears to be *cognizer-first*. This means that an initial acknowledgment or denial of MS's status as a cognizer modulates, and at times determines, whether we think that P is cognitive. To make this thesis more plausible, it is helpful to first consider a case with human subjects. Imagine a human subject (HS) who has suffered extensive brain damage in a traffic accident. Unaware of this fact, we attempt to make sense of his behavior, but whatever process P determines the causal relation between the HS's sensory input and behavioral output, there is no match with functional profiles from our folk-psychological repertoire. Prolonged observation of HS's behavior confirms this picture. If claim (1) is true and we decide whether P is cognitive based on our familiarity with the characteristic functional roles of cognitive processes in guiding behavior, then it appears that we are forced to accept that the process that drives HS's behavior is not cognitive. Moreover, given the findings of our prolonged observation, we would have to acknowledge that HS is not a cognizer.

However, both these conclusions collide with the way in which we attribute cognitive status. Instead of maintaining that noncognitive processing causes HS's behavior, it is much more plausible to think that both the "folk" and the experts caring for HS would say that HS's behavior is caused by *impaired* or dysfunctional cognitive processes. A part of why they are dysfunctional *cognitive* processes is exactly because they belong to a cognizer yet fail to causally connect sensory input and behavioral output in a way that is consistent with our repertoire of functional profiles.

One could object that this result is only yielded in thought experiments involving human beings, but it is plausible that a similar result can be reached with Martians. Imagine that upon intense observation, we find that our Martians display behavior that conforms with our functional profiles (e.g., when they are pinched, they exhibit pain response behavior, etc.), which leads us to conclude that there is cognitive processing. The Martians are thus genuine cognizers. There is, however, one MS whose behavioral patterns are hard to make sense of (when pinched, MS does not manifest any avoidant pain behavior, but smiles, etc.). If we think that coarse-grained functional similarity is a necessary condition for an organism to qualify as a cognizer, then we must conclude that MS is a noncognizer. Nevertheless, in part because of our antianthropocentric leanings, which Clark rightly notes, it strikes me as more plausible that we would instead say that MS is undergoing a form of "mad pain"[14] that does not occupy the usual functional role. If this alternative is more plausible, which I think it is, then it yields the conclusion that, at least sometimes, our approach is *cognizer-first*: we attribute cognizer status to MS, without MS's relevant processes exhibiting coarse-grained functional similarities.[15]

What should our conclusion be here? The two cases with HS and MS were designed to make plausible that our approach to attributing cognitive status is, at least sometimes, *cognizer-first*. Cognitive states are sometimes attributed to subjects based not on whether they provide an explanation of the subject's behavior that matches our functional profiles, but on whether the subject is seen *as belonging to the kind "bearer of cognitive states."* If this is right, then claim (1) is false: whether a process P qualifies as cognitive does not, or at least not entirely, hinge on P's occupying a distinctive functional role that matches our folk-psychological repertoire of coarse-grained functional profiles. Thus, though EMH supporters are right that fine-grained functional similarities are not necessary for cognition, one important upshot is that neither are coarse-grained similarities. But in that case, the EMH needs an independently plausible argument to demonstrate that the folk knowledge of coarse-grained functional profiles demarcates the realm of the cognitive.

Let us sum up. We may agree with Clark that we possess an implicit and coarse-grained folk-psychological knowledge of the domain of the cognitive but resist Clark's claim that this knowledge is "rooted" in our knowledge of the characteristic functional roles. If it is accurate that at least in some cases we do not attribute cognitive status on the background of our implicit knowledge of coarse-grained functional profiles, then some important consequences ensue. EMH supporters have argued that because process P's exhibiting fine-grained characteristics is not necessary for P to qualify as properly cognitive, the fine-grained approach cannot demarcate the realm of the cognitive. I concur. However, by drawing attention to cases of cognizer-first attribution, I have argued that neither fine-grained nor coarse-grained characteristics are necessary for the attribution of cognitive processing. Making explicit our implicit knowledge of coarse-grained functional profiles can be of great assistance, but it is not enough to specify the boundary between cognitive and noncognitive processing.

4.3.2 Martian-Friendly Grain Parameter

Nevertheless, even if we assume that claim (1) is false, and even if it is true that neither fine-grained nor coarse-grained characteristics demarcate the realm of the cognitive, the EMH supporter could still resort to claim (2) to argue in favor of a coarse-grained approach. More precisely, the point is that the Martian thought experiment could deliver independently plausible reasons for adopting a view of cognition compatible with a grain parameter set sufficiently coarse to allow for intelligent Martians to qualify as cognizers. However, if claim (1) is false, then it would at least overshadow claim (2), as it no longer can be taken for granted that we attribute cognition to Martians *because* of the match with cognitive profiles from our folk-psychological repertoire. But independent of the status of claim (1), it is not difficult to see that the success of the argumentative strategy in favor of claim (2) essentially depends on the thought experiment. But just how convincing is it?[16]

Thought experiments gather intuitional data for philosophical theories and can elicit "strong beliefs" in those exposed to them (Parfit 1984), and sometimes they constitute "arguments about counterfactual possibilities" (Williamson 2007, 207). An intuition is typically treated as yielding appropriate data and offering prima facie support for theories that predict such intuitions.[17] However, just as with empirical experiments, the consistency of the premises of the thought experiment, the rigor of the research design, the clarity of background conditions, and the thoroughness with which the conclusions are derived are decisive for the epistemic status of the results. To minimize the risk of contaminating the results, the experimenter has to ensure *internal validity* (correct measurement of the effect between independent and dependent variables) and *external validity* (the results can be generalized to other situations). Let us now explore whether the Martian experiment faces challenges on these fronts.

Regarding external validity, thought experiments are occasionally deficient, in part because judgments suitable in the world of the thought experiment might be contingent on a normative background context that is specific for that world (Wilson 2016; see also Elster 2011). Moreover, in many cases, it is dubious to assume that intuitions elicited reflect universal norms, because consensus on the evoked intuition necessitates something like a common cultural and even educational background.[18] Although such limitations should be kept in mind, the Martian experiment under investigation appears much less intricate in this regard because it does not implicate the type of ethical intuitions and judgments that might plausibly depend on particular normative contexts.

Regarding internal validity, on which much of the epistemic force of the experiments depends, the situation is far more complicated. Intuition is used both as data for developing a theory about the nature of cognition and as a premise in the argument for the coarse-grained view. But is the design of the Martian experiment such that it allows us to make correct judgements about the hypothesis that it aims to test? In the following, the aim is to unearth and discuss two problems arising from methodological considerations. It will be argued that the experiment suffers from shortcomings related to intelligent *Martians* (section 4.3.2.1) and *intelligent* Martians (section 4.3.2.2).

4.3.2.1 Intelligent *Martians*? One of the most severe challenges that thought experiments face is *methodological underdetermination*, which occurs when experimenters fail to provide comprehensive information about background conditions, controlled variables, and possible tests of validity (see K. V. Wilkes 1988). All thought experiments must address this general challenge, but the task is particularly difficult in "outlandish" sci-fi experiments that are remote from everyday life.[19] Even in nonoutlandish thought experiments, a concern is that the experiment is successful "because the evoked response becomes dominant, so that the subject comes (either reflectively or unreflectively) to represent relevant non-thought experimental content in light of the thought experimental conclusion" (Szabo Gendler 2007, 86). Although this concern is less acute in thought experiments involving simple, ordinary entities and scenarios, it is more critical when the reader is invited to imagine Martians and other extraterrestrial creatures. It is particularly in such cases that underdetermination can have detrimental effects, leading to uncertainties about what intuitions are being solicited.

In our case, a central issue is that underdetermination leads to a lack of sufficient controllability. We are invited to imagine Martians, but we lack an appropriate description of both details and background conditions, which makes it difficult to control the parameters by which one's imagination is set to work and to differentiate the activated concepts and schemas (see K. V. Wilkes 1988).[20] Without further specification and without a specific test of validity offering control parameters, it is difficult to ascertain that the crucial apparatus of the experiment, imagination, is being put to use correctly. This means that without a test, we have no means of securing that the Martians are *correctly* imagined, which should give us at least a preliminary reason to worry about the conclusion. Of course, accepting that underdetermination is a general problem, and a particularly serious one in "outlandish" thought experiments, does not warrant a wholesale rejection of the Martian experiment. But it gives rise to legitimate concerns, which motivate further investigation into possible ways in which Martians may be imagined.

The point is that lacking a test of validity, there is no straightforward way of excluding the possibility that when we engage in the thought experiment, we actually end up *imagining the Martians as humanoid*. In fact, this scenario is rather likely. A brief look at the status of Martians in the Western popular collective imagination reveals that they are predominantly imagined and depicted as humanoids. Since scientists observed a network of lines on the surface of Mars in the late nineteenth century, the idea of Martians as a humanoid monoculture has attained an important place in popular collective imagination. Not only does this idea have a long history in our cultural imagination, it has a strong presence that extends from nineteenth-century novels to mid-twentieth-century sitcoms and contemporary Hollywood blockbusters.[21] There are myriad ways in which humanoid Martians are depicted, but they typically embody different traits that we commonly attribute to human nature: some exemplify our vision of our species as being on a peaceful mission, whereas others as embody our (self-) destructive side, colonizing and enslaving entire populations.

Given that it is deeply ingrained in Western cultural imagination that Martians are humanoid, it seems reasonable to expect that the Martians in folk psychological intuitions will not be radically different. To be clear, the point is not that this is necessarily the case, but only that it is very likely that our imagined Martians are humanoid. But in that case, and given the lack of a test of validity, it becomes close to impossible to ensure that the thought experiment solicits the right intuitions. If our imagined Martians are sufficiently humanoid, we may very well end up activating nonrelevant folk-psychological intuitions *about human beings* (who perhaps exhibit a radically different physiological makeup) or perhaps *Martianoid human beings*. But then the experiment would lack sufficient validity, leading to the conclusion that the intuitions solicited cannot function as appropriate support for a theory. Without excluding, or at least making it improbable, that we imagine humanoid Martians, it is not clear what the thought experimenter is entitled to conclude. Clearly, the fact that the experiment activates intuitions that support the EMH cannot serve as evidence that one has imagined the Martians sufficiently correctly to have valid intuitions.

4.3.2.2 *Intelligent* Martians? Related to concerns about validity, there is another aspect of the Martian thought experiment that merits attention.

To see this, let us shift our focus from "Martians" to "intelligent." It is not hard to see why the experiment is explicitly limited to *intelligent* Martians. It indeed appears intuitive that if we encountered Martians that manifested intelligent behavior but did not display the fine-grained characteristics of human beings, we would not hesitate to grant them cognizer status. Nonetheless, there is something puzzling about the instruction to limit our imagination to intelligent Martians. The problem is that the concept "intelligence" usually includes reference to skills that are at least partly cognitive. If cashed out in terms of suitable relations between sensory input and behavioural output, then "intelligent" may not mean much else than "cognitively mediated." Let us consider two possible scenarios, however, one in which the thought experiment leads to the activation of the folk-psychological concept of intelligence and one in which it leads to the activation of the scientific concept.

If the thought experiment activates the folk concept of intelligence, then the conclusion may not represent an epistemic gain. This is because the folk concept "intelligence" includes not only the ability to use nonbasic cognitive capacities but some success in academic performance (Hunt 1987, 281), as well as abstract practical and social skills, including intellectual curiosity and sensitivity to other people's needs (Boyle, Stankov, and Cattell 1995, 482). In case this folk concept is activated and the experimenter imagines the Martians as intelligent *in this sense,* then the experiment will likely yield the intuitive result that Martians qualify as cognizers, yet without actually providing any evidence for this result.

If the thought experiment activates the scientific concept of intelligence, the conclusion is similar. For instance, the American Psychological Association operates with a relatively basic understanding of intelligence as the ability to engage in forms of reasoning leading to flexible behavior, to learn, to adapt to the environment, and to solve problems by using thought. But here, too, we reach a similar conclusion. If the scientific concept of intelligence is activated, then the experiment will yield the result that Martians qualify as cognizers, but without this result being based appropriately on evidence.

Issues with underdetermination prevent us from making reliable predictions about which concept of intelligence the thought experiment will activate. Nonetheless, for the reasons mentioned earlier, both possibilities lead to results that have questionable epistemic value. Moreover, it seems that the problem is not isolated to the thought experiment. Clark (2010b, 92–93) defines what counts as cognition with recourse to the notion of intelligence: "What makes a process cognitive ... is that it supports intelligent behaviour." Thus, process P's supporting intelligent behavior is necessary *and sufficient* to qualify P as cognitive. Analogously, subject S's display of intelligent behavior is a necessary and sufficient condition for S to qualify as a cognizer. But then the problem becomes quite apparent: although valid, the argument is vacuous; nothing new is expressed that is not already contained in the central terms. The conclusion that Martians are cognizers is in effect equivalent to the premise that Martians exhibit intelligent behavior (recall that the experiment is limited to *intelligent* Martians) and that "intelligence" (on both accounts) includes reference to cognitive skills.

To sum up, Clark's argumentative strategy to offer support for claim (2) relies on the Martian thought experiment. It aims to persuade the adoption of a view of cognition compatible with a grain parameter set sufficiently coarse to allow for intelligent Martians to qualify as cognizers. In this section, my main concern was to offer methodological considerations regarding thought experiments and to expose two problems related to both the Martian-ness of Martians and their intelligence. The main conclusion is that the thought experiment lacks sufficient validity and fails to offer adequate support for the coarse-grained approach. But without the support of the Martian intuition, Clark's choice of grain has no independent backing. One could object that he sets out to show that cognition sometimes extends, but though his choice of the appropriate (coarse) grain lacks independent support, it already *entails* that cognition sometimes extends.[22]

But even if we assumed that the targets, Martians, are correctly imagined and are somehow rendered immune to the objections raised here, it would still be an open question whether the elicited intuitions are sufficiently reliable. In scientific practice, to ensure reliability, observational data and measuring equipment is calibrated at particular intervals, which sets a base level and ensures conformity to an independently acknowledged standard.[23] But how do we calibrate intuitions about Martians? Some intuitions appear straightforwardly amenable to calibration. For instance, it is possible to calibrate intuitions about the validity of a modus tollens argument form by a process that does not invoke intuitions: the construction of a truth table (Schukraft 2016). Other intuitions can be calibrated by independent calculations or perhaps based on a suitable overlap between independently justified beliefs and the content of a particular intuition. But in our case, it is hard to see how we could calibrate intuitions about Martians in a process that either does not invoke intuitions or at least invokes intuitions that are independently calibrated. The best option would be to rely on intuitions about cognition that have acquired a calibrated status through antecedent and independent tests. But this option is not available. Not only would it be extremely difficult to distinguish between the intuition-calibrating test and our intuitive responses to the test, but also it would not help us reach the goals of our inquiry: the very reason philosophers elicit intuitions about Martians is that we do not have recourse to a theory or even widely shared intuitions about what cognition is.

4.4 Explaining Conflicting Intuitions

In light of our conclusions so far, the Martian experiment fails to deliver independent support for the coarse-grained view. What are the consequences? Trying to avoid a deadlock by motivating claim (2), Clark (2010a, 62) notes that he fails to "see how to further argue this case with anyone whose intuitions differ." If we assume that Clark is right about the crucial importance of the fine-grained versus coarse-grained question, and if I am right about the Martian experiment, then the worry is that we reach an impasse created by clashing intuitions. Clark thinks that the coarse-grained approach is clearly intuitive and consistent with common sense, but critics have opposing intuitions, thinking that the coarse grain parameter and the location of cognitive activity are at odds with common sense.[24]

To explore how some progress could be made on this issue, we need to investigate the reasons for the clash of intuitions. Although "surface" intuitions of laypeople lacking philosophical training are often muddled and yield unclear results (Kauppinen 2007; Sosa 2007), one might suspect a prima facie problem if rigorously trained professional philosophers eliciting properly directed intuitions about a circumscribed subject fail to reach at least some consensus (Sosa 2007). Certainly, the presence of clashing intuitions should make us cautious in our applications of the relevant philosophically significant concepts (see Williamson 2004). But, we may add, at least in cases where a deadlock threatens, it should also motivate us to determine its source and possible implications for our inquiry. Acknowledging that there might be several possibilities,[25] the focus will be on the structure of the concept "cognition."

4.4.1 Prototype Concepts

One could argue that clashing intuitions are not overly surprising: the intuitions elicited might be muddled by the inherent ambiguities in the concept of "cognition." More precisely, when appealing to folk-psychological intuitions about cognition, we may not be activating a concept with necessary and sufficient conditions for category membership. Rather, we might be activating a *prototype concept* with a probabilistic structure, in which membership in the extension of the concept is graded. A prototype theory of concepts (Rosch 1973; Rosch and Mervis 1975)[26] has productively been deployed to reconceptualize key concepts in psychology, such as intelligence, self, personality disposition, emotion, and concepts.[27] Such reconceptualization has helped both to account for a variety of psychological phenomena and to provide an explanation for why researchers have been unable to produce definitions in terms of necessary and sufficient conditions.

From such a background, it is plausible to assume that, rather than being classically defined, "cognition" is organized around prototypes. Not only have researchers failed to provide necessary and sufficient conditions, but cognition also subsumes a much larger number of individual psychological and neural processes than other concepts in psychology that are now regarded as having a prototype structure. According to a traditional account, cognition includes but is not limited to the storage, transformation, recovery, and use of external or internal input related to language, memory, attention, decision-making, problem-solving, imagination, perception, planning, and so on (see Neisser 1967). Moreover, the field is moving toward an even broader understanding—for instance, questioning the ontological boundary between cognition and affect (Duncan and Barrett 2007).

If cognition has a prototype structure, then whether something qualifies as *cognitive* is a matter of degree and hinges on grades of similarity to prototypical cases. Actual occurrences of cognition then can be ordered by their degree of membership, but no single feature is necessary for something to fall under it. Prototypical cognitive processes that the cognitive kind is centered on gradually shade into nonprototypes, which again gradually shade into nonmembers. Moreover, it is not only that marginal nonprototypes have few or no attributes in common; in many cases, they exhibit attributes in common with members of neighboring concepts. Further muddling the picture, just as cognition is prototypically organized, the kinds that it subsumes (memory, problem-solving, etc.) may each be prototypically organized. In that case, "cognition" would indeed conceal something like a conceptual minefield.

It appears plausible to assume that cognition exhibits a prototype structure, and it also offers a possible way to make sense of the clashing intuitions of philosophers. First, because of the lack of a neat assortment of properties possessed by all members, there will be examples that fail to capture all intuitions. Certain intuitions will only make sense when applied to prototypical cases but not to peripheral ones. The vagueness generated by this structure would allow that process P might be a genuine borderline case, in which case no amount of inquiry could settle whether P is cognitive. In other words, trying to decide whether P is really cognitive may in some cases turn out to involve challenges similar to trying to decide whether gray is actually black or white. But then, the existence of two conflicting intuitions held by properly trained philosophers does not necessarily mean that one of them is wrong.

Our opponent may grant the existence of cases that fail to capture all intuitions, as well as the existence of genuine boundary cases, but still reasonably doubt that this explains the profound kind of impasse that we are investigating. To address this concern, and to render the clash of intuitions between philosophers less perplexing, we need to dig deeper into the structure of the concept. The proposal is that "cognition" is stretched out between relatively disconnected focal quasi-prototypes, Q_1, … ,Q_n, that loosely link convex sets in different locations. Cognition is then a placeholder for several focal quasi-prototypes. To see how reconceptualising cognition as exhibiting a prototype structure can help explain the clashing intuitions of philosophers, consider three additional points.

First, the focal quasi-prototypes could be located on different levels. Cognition features in true statements on a number of levels involving different types of properties. It features in generalizations on the coarse-grained level of behavior (patterns of action), psychology (beliefs, desires, etc.), and in generalizations on the fine-grained level of vehicles (mechanisms driving cognition), neural activity (activated brain areas), and molecular and cellular activity (kinase activation, synaptic plasticity, etc.). Although generalizations from all these levels do not straightforwardly align, intuitions of different philosophers may result from, or at least be significantly influenced by, the implicit activation of focal quasi-prototypes located on these different levels, resulting in different judgments about the classes of organisms that qualify as cognizers.

Second, the focal quasi-prototypes could be located on the same level. For instance, we may understand the structure of the concept as organized around four focal quasi-prototypes, Q, which are located on two continuums, (Q_1, … , Q_2) and (Q_3, … , Q_4). The first connects fine-grained quasi-prototype Q_1 (e.g., a particular case of memory exhibiting recency and primacy effects) to a coarse-grained Q_2 (e.g., the case of Otto, artificial memory). The second connects Q_3 (a conscious occurring thought) and Q_4 (a nonconscious process, perhaps in long-term memory). When trying to arbitrate whether a process belongs to the cognitive kind, the activation of different, but same-level, focal quasi-prototypes may lead to dissimilar intuitions.

Third, prototype concepts allow for and often involve considerable *prototype overlap*, which refers to the fact that an instance can simultaneously be a peripheral member of one concept and a more prototypical member of another, neighboring concept. Thus, when eliciting intuitions, the salience of some features may be diminished or annulled by the possession of other features more directly suitable to a neighboring concept (for a discussion, see Ramsey 1992). In our case, this neighboring concept is "mind." Not only are there significant overlaps between these concepts, but in the EMH literature, *cognitive* and *mental* are almost without exception used interchangeably. This conceals the fact that the fine-grained versus coarse-grained question itself echoes a broader debate about cognition and mind. This broader debate involves discussions between classical computationalists (maintaining that cognitive processes are basically computations) and connectionists (cognitive processes are algorithms for activation in a network),[28] but also longstanding quarrels about the permissibility of attributing cognitive predicates like "knows" and "realizes" to philosophical zombies, computers, blockheads, thermostats, swampmen, Pollock's OSCAR system, and so on. The point is that though such an overlap with the neighboring concept "mind" is extremely hard to get a grip on, it may be causally implicated in the generation of conflicting intuitions.

To sum up, the claim that cognition exhibits a prototype structure is plausible in light of the lack of a definition of the concept, the extensive area of processes that it covers, and evidence from psychology about classical versus prototype concepts. While conceding that more evidence is required, we may note that thinking of "cognitive" as stretched out between several focal quasi-prototypes yields epistemic gains. It allows for examples that fail to capture all intuitions and helps us unearth some of the complexities that can influence intuitions, but also allows us to explain how two trained philosophers may elicit conflicting intuitions, without being forced to maintain that one of them is necessarily wrong. They might be guided by less prototypical kind members that do not share significant characteristics. Along these lines, such disagreements could be viewed as more like theoretical disagreements in the sciences that have accepted pluralism—a topic that is further considered in the next section.

4.5 A Plea for Pluralism

In light of clashing intuitions, profound disagreements about the realm of the cognitive, and the analysis offered so far, what should our conclusion be? Three possibilities jump to mind:

(1) *Optimism:* Many implicitly assume that a single, unified framework for all of cognition is possible, despite the complexity of its subject matter. In fact, some argue that the emergence of embodied approaches will itself help generate unification. As Glenberg, Witt, and Metcalf (2013, 573) note, "Having the body in action as a central consideration for theories of cognition promises, we believe, to help unify psychology."

(2) *Eliminativism:* "Cognition" not only fails to form a natural kind, but is structured around several prototypes, which is why it is too ambiguous to function as a proper object of scientific inquiry. Embracing what we could call an eliminativist position, one might argue that "cognition" fails to map onto a kind with sharp boundaries and should therefore be eliminated from scientific taxonomy.

(3) *Pluralism:* The patterns of interaction between brain, body, and environment that constitute cognitive processes are probably sufficiently

> complex to admit several legitimate grain levels of analysis. Given such complexity, it is unlikely that any approach based on one single grain level will be able to satisfy the entire range of our scientific interests, including, but not limited to, explanatory needs, inductive generalizations, and predictive utility.

One might be tempted to reach the radical conclusion that "cognition" not only fails to form a natural kind but also is structured around several quasi-prototypes and hence too ambiguous to function as a proper object of scientific inquiry. Yet, before opting for conclusion (2) and eliminating "cognition" from scientific taxonomy, it is worth calling to mind that it is usually prudent for philosophers to exhibit moderation when articulating normative proposals about taxonomic changes in science. Following this defeasible rule, normative changes advocated by philosophers at least should be consistent with established scientific practices that are generally recognized as successful (Machery 2009, 237).

To make an informed choice from among the alternatives, we may seek inspiration in scientific fields that are established and generally recognized as successful. Ideally, ensuring adherence to this rule would require a systematic investigation of randomly selected samples drawn from many relevant cases from different disciplines. But for our purposes, it is enough to pick out a few examples showing how mature sciences like biology have dealt with similar issues. For instance, it is widely acknowledged that despite long and intense debates, biology has not been able to demarcate the realm of "life" and has failed to provide a single generally accepted characterization of "genes" (Sterelny and Griffiths 2010; Griffiths and Stotz 2006). Those who describe their research as investigating the key molecular constituents of life (microchemical systems capable of undergoing Darwinian evolution or RNA molecules able to self-replicate and evolve) tend to operate within a fine-grained concept of life. In contrast, those who regard the entire biosphere as alive or claim that whether an entity is alive is substrate-independent and hinges instead on organizational structure tend to operate within a coarse-grained concept of life. In a similar manner, scientists portraying their research as examining fundamental constituents of genes (microscopic mapping of the DNA molecule) tend to operate within a fine-grained concept of gene, unlike scientists who provide purely informational descriptions of the genome. Importantly, this has not led scientists to embrace eliminativism, but rather to accept pluralism about life and gene (see Griffiths and Stotz 2006; Sterelny and Griffiths 2010). For instance, differentiating three senses of *gene*, Griffiths and Stotz (2006, 499) maintain that "DNA sequences that are genes in one legitimate sense of the term may not be genes in another equally legitimate sense."[29]

From this background, we may start to think that pluralism about cognition could be an appealing alternative, particularly in light of the fact that neither proponents of embodied cognition (EC) nor cognitivists insist on being able to explain the entire range of cognitive processes. Prominent proponents of cognitivism now emphasize that computational theory can only provide a part of the story about cognition. Looking back on his early work a quarter of a century after *The Language of Thought*, Fodor notes that the computational theory was not meant to suppose "that it could comprise more than a fragment of a full and satisfactory cognitive psychology ... the most interesting—certainly the hardest—problems about thinking are unlikely to be much illuminated by any kind of computational theory we are now able to imagine" (Fodor 2000, 1).

Others have taken cautious steps toward accepting some partial form of pluralism. For instance, Wheeler (2005, 111) rejects the idea that "the class of computational systems and the class of dynamical systems don't intersect at all" because computational systems are symbolic systems. Although he holds onto the idea that representation is necessary for computation and that a computational system realizes distinct input-output functions by manipulating representations, in his "liberal" approach he understands computational systems as a subset of dynamic systems (Wheeler 2005, 115 and 298). Although much hinges on the claim that dynamic systems explanations appeal to representations, the liberal framework allows for hybrid explanatory models that encompass both computational and noncomputational processes.

The pluralist approach suggested here goes beyond the liberal approach. It distances itself from the view that the cognitive status of processes and states crucially depends on the presence of representations. Instead, following Ramsey (2017), it remains neutral on the significance of representations as explanatory conjectures while rejecting the notion of representation as a demarcation criterion. As Ramsey (2017, 4201–4202) puts it, "Our tendency to define cognition in representational terms or to simply equate mental processes with representational processes stems from presuppositions that are not well-founded and that are prone to hinder scientific theorizing." Importantly, though the pluralist approach suggested here does not require appeal to representations to form hybrid models, it does not oppose the idea that particular cognitive tasks require representations. Rather, it aims to take seriously the notion that it is not possible to offer a single account of cognitive phenomena constructed on a single set of principles that make recourse to representations. Moreover, in this view, the sheer diversity of extant approaches in cognitive science is not necessarily a sign of deficiency and is consistent with the view that there is no uniquely correct description of the mechanistic structure of the world (Craver 2009). Pluralism would allow for a variety of approaches that aim at different degrees of grain level, specificity, and abstraction, and it would reflect the diversity and complexity of the structure of the world.[30]

4.6 Some Objections to Pluralism

It is perhaps helpful to further explicate the idea behind the pluralist approach suggested here by showing how it responds to two concerns. Opponents might worry that (a) pluralism implies antirealism and/or that (b) pluralism impedes scientific progress. In reply to concern (a), it is important to stress that pluralism merely acknowledges that nature has innumerable joints, which means that there can be several legitimate ways of carving it. This, however, does not imply a commitment to antirealism. It merely rejects that the structure of the world dictates a forced choice between fine- and coarse-grained approaches and, for that matter, between representational and nonrepresentational approaches. In fact, this sort of pluralism is consistent with sophisticated versions of realism that respect the relativity of grain parameters to scientific contexts of investigation (see Chakravartty 2011). Consequently, though the pluralist approach suggested here denies that there is a single privileged approach, it does not claim that the ultimate understanding of the world is achieved by an unrestrained proliferation of different perspectives. It does not deny the realist commitment to the existence of a mind-independent reality, nor the understanding of theories as supplying knowledge of observables and unobservables. It is in this sense a pluralist-realist view (see, e.g., Mitchell 2009).

Instead of a bottom-level reductive explanation, pluralism makes room for integrating several models and explanations at several levels of analysis, acknowledging that there are several ways to accurately represent parts of the world, with different levels of generality and abstraction. The suggestion is not that there are multiple worlds, but that it is possible to individuate processes and objects in several ways, such that they reflect both causal structures of the world and explanatory interests.

In reply to concern (b), it is important to show that pluralism about cognition does not mean that the concept of cognition is not useful and should be replaced with several concepts that denote "real" categories. The concern may be largely neutralized, once again, by considering some scientific areas that have adopted pluralism. As Sterelny and Griffiths (2010, 355) have argued, not only is pluralism about life and genes accepted in biology, it is also the case that pluralism has not impeded scientific progress. Others argue that pluralist competition in chemistry has, in fact, propelled the development of productive theoretical approaches (see, e.g., Chang 2012). If this is true, then there is at least nothing inherently obstructive in adopting pluralism. At the very least, this shifts the burden of proof to those who think that the case of adopting pluralism in cognitive science would play out differently. In a similar way, though science is not obligated to preserve folk concepts, opting for elimination over pluralism would require evidence that the usage of an inclusive concept of cognition is disadvantageous. But such evidence is currently lacking. Worse, one may even suspect that in some cases, it is the resistance to pluralism that obstructs advancement. Discussing the nature of representations, Ramsey (2017) notes that the lack of pluralism impedes progress. Although he is no supporter of the idea of bolder forms of scaffolded cognition, he argues that cognitive science should not artificially limit theory development by requiring explanatory posits like representations. Ramsey (2017, 16) thus argues against the "common practice of demarcating cognition and cognitive theorizing in representational terms." But if Ramsey is right in maintaining that the representational versus nonrepresentational divide does not demarcate cognition, then we have at least initial reasons for favoring a pluralist approach.

Having offered reasons to neutralize some concerns about pluralism, it is useful to complement the picture with the additional point that pluralism seems an obvious choice given the complexity of cognitive science's subject matter range, its interdisciplinary domains of inquiry, and its being a young science (see Dale 2008; Dale, Dietrich, and Chemero 2009). When dealing with complex domains, carving up the world in a pluralistic way (that might prevent integration into one larger picture) may often be the most constructive approach (Horst 2007). Instead of somewhat unproductive boundary debates, it is more helpful to accept that, implicitly or explicitly, research projects with different explanatory interests use different grain parameters and imply the existence of different scientific entities.

The pluralist approach suggested here does not deny that a single account could generate adequate explanations for an extensive range of cognitive phenomena. However, it accepts that in some cases, it is possible to arrive at different, but valid, explanations, whereas in other cases, the integration of different explanations leads to the best explanation. For instance, Smolensky and Legendre (2006) have successfully combined connectionist and symbolic architectures, suggesting the possibility of theoretical integration for other cognitive processes. Other cases show that embodiment and cognitivism might not be mutually exclusive competitors but might provide a complimentary explanation. A closer investigation of the CajunBot robotic device, which can autonomously navigate over varying terrains, finds the use of mixed strategies—some of them best described in cognitivist terms, others in embodied terms (Berkeley 2008, 96). So perhaps ASC and cognitivism are best conceived of not as competitors, but rather, at least in some cases, as complementary approaches.

For a brief example, consider how visual perception might be an area that requires a pluralist approach. The *two-streams hypothesis* divides the visual stream into a ventral stream (or *vision-for-perception* pathway, recognizing and discriminating entities and shapes) and a dorsal stream (or *vision-for-action* pathway, reaching and grasping guided by vision; Milner and Goodale 1995). Given this structure, some argue that investigating the pathways of vision requires both representational and nonrepresentational approaches. For example, Norman (2002, 89) argues that "both approaches to perception ... are valid descriptions of perception, but of different aspects" and calls for an integration of the two approaches into a single, dual-process approach. The idea is that a representational account is best suited for the ventral stream, whereas a nonrepresentational account is appropriate for the dorsal stream.

4.7 Conclusion

The debate between cognitivism and EC, as well as between EC accounts, often revolves around the question of the boundaries (the "where") of cognition, which is intertwined with the question of ontological inseparability. This chapter first engaged and tested two accounts of the mark of the cognitive that distinguish mere information processing from cognitive processing, followed by an investigation of the fine-grained versus coarse-grained debate about the question of the correct grain level that should be deployed to examine cognitive processes. The chapter dealt with an important argument for setting the benchmark for parity by recourse to coarse-grained functional profiles, which relies on a certain view of the folk-psychological knowledge of the cognitive domain, and on a thought experiment involving intelligent Martians. We concluded that due to important shortcomings, this argumentative strategy fails to offer adequate support for the coarse-grained approach.

As the debate leads to an unproductive impasse, the chapter explored two underlying sources of the opposing intuitions held by philosophers. It was argued that the structure of the concept "cognition" and the likely presence of systematic biases could be seen as major contributing factors. With a better grip on the nature and sources of the clashing intuitions, the last part of the chapter argued that elimination from scientific taxonomy is unwarranted. Instead, the disagreements about the boundaries of cognition that cut across much of the philosophy of cognitive science were here regarded as an opportunity to explore the possibility for a kind of nonthreatening pluralism that is familiar from mature sciences. Expert disagreements were made intelligible by recourse to prototype concepts, with several prototypes pulling toward different standards of application. This does not mean that the disagreement is merely a verbal one, nor that "being cognitive" is a graded property like "being wet." The point is that instead of looking for a clear boundary of demarcation, we should accept that there might be no such thing as a concept of cognition that is optimal across the entire variety of contexts. This is particularly attractive if we accept that it would be erroneous to invoke representation as the common aspect involved in all instances of cognition.

Although the reasons for accepting pluralism appear convincing, and consistent with successful and established scientific practices, much will depend on the theoretical and empirical benefits that it might help accumulate in domains of scientific research. Making the case for a pluralist approach is also consistent with the fact that embodied approaches have had successes, especially in explaining low-level perception, action, motor control, and coordination, but less so with explaining cognitive skills that require stepping back from interacting with the world and thinking carefully (Chemero 2013). We have reasons to think that no single account of cognition is both broad and specific enough to account for a wide variety of cognitive processes.

5 Scaffolding and Dependence

5.1 Notions of Dependence

The notion of scaffolding can be utilized productively to consider a variety of relationships of dependence of cognition on the body and the environment. Nonetheless, the flip side of this flexibility is that scaffolding depicts too wide a variety of relationships of dependence.[1] In this admissive sense, whenever the functioning of Y depends on X in some form, then X scaffolds cognitive process Y. The introduction of the notion of active scaffolding excludes trivial forms of dependence, denoting a particular relationship of *nature-dependence* (the scaffolded depends on the scaffold not merely for its existence, but for its nature or character) and *difference-making* (it is possible to manipulate the scaffolded by manipulating the scaffold).

We are now able to further specify the exact nature of the metaphysical relationship. In its current form, active scaffolding is still too unspecific, picking out relationships of dependence that differ from both metaphysical and practical viewpoints. The relationship of dependence between X and Y is consistent with two distinct metaphysical relations. Y's being actively scaffolded onto X could be comprehended in terms of a *relation of causation* or a *relation of constitution*. To further specify the relation of dependence that ASC aims to pick out, this chapter will seek alternatives to the debates discussed in chapter 4 by seeking guidance from explanatory practices in the relevant scientific fields.[2] Recent work on mechanisms and mechanistic explanation (Machamer, Darden, and Craver 2000; Craver 2007) offers valuable tools to rethink causal and constitution relations of dependence that could further specify the types of relationships that ASC gestures toward and to help distinguish genuine components from causal background conditions. Since the idea of using the literature on mechanistic explanation to propel debates in the philosophy of cognitive science was first proposed (Kaplan 2012; Zednik 2011), it has attracted the attention of several philosophers (e.g., van Eck and de Jong 2016; Kirchhoff 2017; van Eck 2018; Krickel 2018; Baumgartner and Wilutzky 2017). In addition, the use of an interventionist account of causation to assist in determining relationships of mutual manipulability between putative components offers several advantages. First, it provides a basis for detailing ASC, which avoids the risks discussed in chapter 4 and offers more neutral guidance than intuitions about grain level. Second, it dovetails with the general pragmatic ambitions of this book that seek to explore the possibilities of extending the range of therapeutic interventions. Explanations in cognitive science and neuroscience often are driven by the aim to develop new ways to intervene in neural and cognitive activity—for example, to prevent and treat illnesses. Third, it retains proximity to explanatory practices in the relevant scientific fields and offers considerations on amenability to experimental testing and confirmation.

The notion "active" and talk of "difference-making" can be understood as suggesting some kind of causal relationship between the scaffolding X and the scaffolded process Y. This chapter starts by exploring this possibility, understanding causal relationships along the lines of an *manipulationist* or *interventionist approach* (M) that comprehends causal relations in terms of being utilizable for manipulation (Woodward 2003, 2010, 2015). Adding auxiliary conditions will further limit the scope and specify active scaffolding, offering what we could call a manipulability account of active scaffolding, or *Active Scaffolding (M)*. There is, however, a form of active scaffolding that is characterized by a more complex mutual difference-making, indicating constitutive relevance. To capture this form of active-scaffolding relationship, this chapter will then turn to the literature on mechanistic explanation and to Carl Craver's (2007) account of mutual manipulability (MM) and offer a mutual manipulability account of active scaffolding, or *Active Scaffolding (MM)*. One of the crucial challenges herein is to avoid importing some of the difficulties that result from using the interventionist approach to evaluate both causal and constitutive relationships.

5.2 The Interventionist Account of Causation

Philosophical accounts of causation subscribe to the view that a cause must make a difference to its effects. This can be explicated in terms of conditional probabilities (given appropriate background conditions, if X → Y, then X must raise the probability of Y) or in terms of counterfactuals that depict how Y would change under interventions on X that alter X. An influential version of the latter approach, which is commonly assumed to underlie research across the sciences, understands causal relations in terms of relations between variables that can take on different values and causes as potentially exploitable means for manipulating effects.

According to such a *manipulability* or *interventionist* account, unlike correlational relationships, causal relationships can potentially be used for purposes of control and manipulation (Woodward 2003, 2010, 2015). A causal relationship holding between *variables* (roughly a property or quantity that can, as a minimum, take on two values) X and Y is thus a matter of counterfactual dependence between X and Y. Claiming that X causes Y (represented as X → Y) means that "for at least some individuals, there is a possible manipulation of some value of X that they possess, which, given the appropriate conditions … , will change the value or probability distribution of Y for those individuals" (Woodward 2003, 40). In a more recent paper, Woodward offers the following formulation: "M X causes Y if and only if there are background circumstances B such that if some (single) intervention that changes the value of X (and no other variable) were to occur in B, then Y or the probability distribution of Y would change" (Woodward 2010, 290).

An additional distinction between contributing and direct causes helps unearth the number of possible relationships that share the common structure of being tied to the way in which Y behaves under an appropriate intervention on X while other variables are kept constant. X is the direct cause of Y just in case there is an intervention on X with respect to Y, which will alter Y (or the probability distribution of Y) while all other variables are held fixed. X is the contributing cause of Y just in case there is a directed path from X to Y (with a direct causal relationship between each link) and there is a possible intervention on X with respect to Y that alters Y while all other off-path variables are kept fixed (Woodward 2003, 59). In this manner, M explicates what it is for a causal connection to hold between X and Y in terms of the results of an ideal experimental intervention I on X with respect to Y.[3] We may represent this as follows:

I on X → Y

Interventions thus involve variables, the values of which are determined from outside the system (not necessarily by an experimenter). Woodward (2015, 311) maintains that an intervention is "an idealized, unconfounded experimental manipulation which would be appropriate for determining whether one variable is causally related to a second variable." I is an intervention variable for X with respect to Y if and only if I meets four conditions:

(1) I causes X.

(2) I acts as a switch for all the other variables that cause X. That is, certain values of I are such that when I attains those values, X ceases to depend on the values of other variables that cause X and instead depends only on the value taken by I.

(3) Any directed path from I to Y goes through X. That is, I does not directly cause Y and is not a cause of any causes of Y that are distinct from X—except, of course, for those causes of Y, if any, that are built into the I-X-Y connection itself; that is, except for (a) any causes of Y that are effects of X (i.e., variables that are causally between X and Y) and (b) any causes of Y that are between I and X and have no effect on Y independently of X.

(4) I is (statistically) independent of any variable Z that causes Y and that is on a directed path that does not go through X. (Woodward 2003, 98)

In part, the appeal of M is that it explains why theoretical interests are not satisfied simply by knowledge of correlations and offers a clear distinction between causal and merely correlational relationships, which corresponds to the way such relationships are understood in the sciences.[4] Although X being the cause of Y implies that manipulating X is a way of manipulating Y, X being merely correlated with Y fails to imply that manipulating X is a way of manipulating Y.[5]

To take an example, the *New England Journal of Medicine* published an amusing and lighthearted article that found a correlation between the chocolate consumption of a country and the cognitive performance of its population, measured by the total number of Nobel laureates per capita

(Messerli 2012). Switzerland was found to perform best in both chocolate consumption and the quantity of Nobel laureates. Some media outlets interpreted the results to show that the correlation is due to chocolate consumption somehow causally affecting the number of Nobel laureates. In any case, M is directly relevant to address such questions, and it offers a way to probe the alleged causal relation by experimental manipulation, consisting of an intervention that manipulates the putative cause and observes whether an effect results. The question whether X → Y is a question about what would ensue for Y if an I were performed on X. How would such an intervention look?

Holding all other circumstances constant, an experimenter could supply a group of randomly chosen countries with an increased amount of chocolate (CH), while supplying the control group with the same amount of placebo chocolate, and observe whether or not a statistically significant difference is produced in the number of Nobel laureates (NL), represented as follows:

I on CH → NL

This intervention meets condition (1) because I determines the level of chocolate consumption. Condition (2) is met because only the experimenter's choice to assign a country to a group (control or experimental) determines whether the country has a high chocolate consumption or not. Condition (3) is met because if assigning a country in the experimental group influences the number of its Nobel laureates, then it does so only by way of the elevated consumption of chocolate. Finally, condition (4) is met so long as I is a randomized experiment in which fixing the value of CH is independent of other variables that might influence NL.

Presumably, the correlation between CH and the number of NL is not exploitable and will not be invariant[6] under attempts to use the former to manipulate the latter. Rather, the correlation probably arises due to some common cause W (e.g., the total wealth of the country) that affects both CH and the number of NL in the following way:[7]

CH ← W → NL

Consider another case. A large study involving more than 250,000 participants examined the consumption of several kinds of beverages with respect to major depression (MD) diagnosis. Although an earlier study found that adults consuming more than a liter of soft drinks per day exhibited roughly

60 percent higher incidence of MD or mental problems (Shi et al. 2010), Guo et al. (2014) found that individuals consuming more than four cups per day were up to 30 percent more likely to develop depression, especially those who drank artificially sweetened diet soda (DS). If we want to determine whether the consumption of DS increases the risk for MD, we can design an intervention in which, for a given population, we intervene on randomly selected individuals, instruct them to consume more than four cups of DS, and observe if they subsequently exhibit an increased incidence of MD:

I on DS $\rightarrow$ MD

The requirement is that the averaged differences in level of DS in each cohort result entirely from I, that I does not affect the incidence for MD by any path that does not go through DS (e.g., by increasing BMI, adding stress, etc.), and that I is not itself influenced by any cause that impacts MD via a path that does not travel through DS. If I meets these conditions and it makes a difference to the incidence of MD, then DS stands in a causal relationship to MD. These conditions should exclude alternative explanation MD $\rightarrow$ DS, according to which individuals with MD crave sweetened beverages and opt for diet beverages because they are simultaneously worried about their weight or perhaps already have existing weight problems. Also, the conditions should exclude the possibility DS $\leftarrow$ W $\rightarrow$ MD, according to which a common cause W (e.g., socioeconomic standing, lifestyle, obesity and poor health, etc.) is responsible for both the high consumption of DS and the increased incidence of MD.

5.2.1 Adding Stability and Causal Specificity

Understanding causal relationships along the lines of M does not require specific commitments about the nature of background conditions or how X changes are linked to Y changes. In this sense, M is relatively basic and unrestricted, too much so to capture the particular relationship of nature-dependence that ASC thus far only gestured at. As we will discuss, some relationships that meet M's requirements do not display characteristic features of exemplary causal relationships. We may add some more conditions that further specify how various interventions on X change Y and offer the prospect of more fine-grained manipulability and control. The idea is that more stable and specific causal relationships are likely to offer more opportunities for manipulation and control than those exhibiting less stable and specific relationships.

The relationship X → Y can be more or less stable, depending on the amount of background circumstances in which it occurs.[8] X → Y is relatively stable if the counterfactual dependence holds under a wide variety of background circumstances. Typically, distal causal relationships with numerous links are less stable. To illustrate what a relatively unstable causal relationship looks like, imagine just a few of the effects somebody's decision to start a family can have. A's decision leads to her purchasing a house she would not otherwise have purchased, hiring a landscaper that she would not otherwise have met, leading to the landscaper marrying A's cousin, who then moved to Norway with him and developed seasonal affective disorder (SAD). Even if in M, A's decision to start a family is causally relevant to the development of her sister's disorder, invoking such a relationship is uninformative. The causal relation is unstable: The associated counterfactual dependence holds under the concrete background circumstances, but would not hold had any of them been just minimally different. Any minimal difference, and A's sister would not have fallen ill with SAD in Norway. Although the relationship between A's decision and A's sister's illness may perhaps qualify as causal, it does not display characteristic features of exemplary causal relationships.

Moreover, X → Y can be more or less specific. In this context, *specificity* refers to the grain level of counterfactual dependencies between X and Y. That is, X → Y is specific if the counterfactual dependencies holding between X and Y are fine-grained. The more the causal mapping between values of X and values of Y approaches a bijection, the more their manipulation enables more precise control over the value of Y (Woodward 2010, 305). Very roughly, X has a more rather than less specific causal effect on Y to the extent that a number of possible states of X (X_1, ... , X_n) and of Y (Y_1, ... , Y_n) map onto each other such that a particular state of X is not associated with many distinct states of Y and the number of distinct states of X mapping onto the same state of Y is not too high. X → Y has a high specificity when intervening on X and enables modulating the state of Y in a fine-grained manner.[9] Conversely, switch-like causation has a low grade of causal specificity. These are cases in which Y has many different possible states (Y_1, ... , Y_n), whereas X only has two possible states ("on" and "off"). Intervening on X and changing its state in either way will change Y—for instance, from Y_1 to Y_2—but the majority of the possible changes in Y's state cannot be realized just by manipulating X. Manipulating X is coarse-grained: one cannot use X for a fine-grained modulation of the state of Y and thus cannot affect whether Y_2 rather than Y_3 is realized.

To illustrate causal relations with high and low specificity, Woodward (2010) considers a radio equipped with a tuning dial and an on/off switch. For the listener to enjoy a particular radio station, the switch must be in the "on" state and the dial must be tuned to a specific frequency. The switch and the dial are specific to their effects to different degrees. The influence of the switch is relatively nonspecific with respect to whether a station is heard, whereas the dial is a relatively specific cause of which station is heard. Manipulating the switch will have an all-or-none effect, whereas manipulating the dial will exert a relatively fine-grained causal influence, taking on many possible values and resulting in differences in the station that is heard.

At least for now, we may say that we conceive of the relationship of nature-dependence that ASC aims at in terms of M, adding that the relevant causal relationships are relatively stable and specific. In the context of ASC, we want counterfactual dependence to hold under a wide variety of background circumstances and to behave more dial-like than switch-like. To elucidate the importance of the latter, consider that though inducing stress would definitely disturb a subject's ability to perform a cognitive task, such an intervention will be relatively nonspecific in its effects on the cognitive task. It will behave more switch-like than dial-like and will thus not offer fine-grained control over the performance of the task while diffusely affecting a number of other variables.

5.2.2 *Active Scaffolding (M)*

M offers a clear distinction between causal and merely correlational relationships, along with conditions for manipulation in causal terms. Although M does not consequently reduce causation to a noncausal vocabulary, the circularity is not vicious and it is not difficult to see that M is suitable to medicine and psychiatry. It sheds light on causal claims made based on experimental practices and establishes a productive link to explanatory models in the sciences. It does not merely offer a tool to distinguish mere correlation and causation, but also fits a fundamental pragmatic objective, which is to causally intervene to treat or prevent disorders (Kendler and

Campbell 2009). It therefore appears productive to think of the actively scaffolded relationship along the lines of M.

Nonetheless, M allows certain relationships to qualify as causal that do not display characteristic features of exemplary causal relationships and are best described as causally relevant background conditions. For this reason, it is beneficial to restrict ASC by adding to M conditions further specifying how Y is changed under different interventions on X. Adding conditions of stability and specificity picks out relationships that offer more opportunities for fine-grained manipulation and control. This helps offer a slightly more precise idea of what active scaffolding amounts to:

Active Scaffolding (M): X is actively scaffolded onto Y if X causes Y in a way that fulfills the conditions of M and the relationship is relatively stable and specific.

This how the scaffolded stands in a dependence and difference-making relationship. However, it is not difficult to see that many of the examples of actively scaffolded processes in chapter 3 were characterized by not just unidirectional but bidirectional difference-making. To approach this aspect, we now turn to the literature on "mechanistic" approaches to explanation, which rely on M as a starting point and can be used to elucidate the kind of mutual difference-making that characterizes a second form of active scaffolding.

5.3 Mutual Manipulability

Mechanistic explanation offers an alternative to explanations that make recourse to a covering law or purely causal approaches. The former explains a phenomenon (Ψ) by demonstrating that it can be predicted based on laws and background conditions, while the latter explains Ψ by demonstrating how it is produced by causal factors. Mechanistic explanation proceeds in a different manner (seen in Craver 2007, 107–108). Etiological mechanistic explanation typically comprehends phenomena in terms of their being caused by a mechanism.[10] In contrast, constitutive mechanistic explanation offers explanation in terms of the behaviors and organization of component entities of the underlying mechanism, which stand in a constitutive relationship to the phenomenon. It is often maintained that explanatory practices in the behavioral and biological sciences are best understood as mechanistic in the latter sense.

For example, explaining spatial memory involves the examination of a phenomenon (e.g., a rat navigating a maze) by discerning the activities and organization of constituents (the rat's brain producing spatial maps; Craver 2007; Bechtel and Richardson 2010). Such explanations do not merely offer box-and-arrow diagrams but disclose the *spatiotemporal organization* of a mechanism (Craver 2007, 139), defined as "a structure performing a function in virtue of its component parts, component operations, and their organization" (Bechtel and Abrahamsen 2005, 423).[11]

Put in more formal terms, in a mechanistic explanation, a phenomenon (Ψ) is explained by the configuration of component entities (X_1, X_2, ... , X_n) and their activities (Φ_1, Φ_2, ... , Φ_N). S comprises entities and activities organized in a particular way so as to implement Ψ, but also parts that are not components of the mechanism responsible for Ψ (Craver 2007).[12] In a characteristic mechanistic model (figure 5.1), the explanandum is a phenomenon (top), typically an activity of the entire mechanism (bottom). In Craver's jargon, X_1's Φ_1-ing stands for X_1 exhibiting behavior Φ_1, whereas Ψ can be comprehended as an input-output relationship obtaining between the values of variables Ψ_{in} and Ψ_{out}, where the inputs include the conditions required for Ψ (Harinen 2014; Craver 2007, 145). For the sake of simplicity, Ψ will be used for the behavior of higher-level entity S and Φ for the behaviors of lower-level entities (X_1, X_2, ... , X_n), such that Φ_1 stands for X_1's Φ-ing.

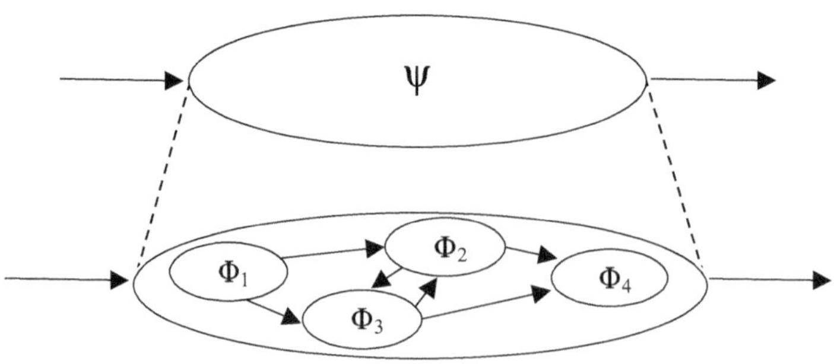

Figure 5.1
A characteristic representation of a mechanistic model. Adapted from Craver 2007, 121.

There is a spatiotemporal overlap between the levels (represented by the dotted line). Constitutive relations are interlevel, whereas causal relations are intralevel (represented by the arrows in figure 5.1; Craver 2007, 121). If X causes Y, then their relationship of dependence is asymmetrical (Y depends on X but not vice versa) and X and Y are mereologically independent entities such that X temporally precedes Y. If X constitutes Y, then their relationship of dependence is symmetrical (Y depends on X and vice versa) and X and Y are spatiotemporally overlapping entities.

Mechanistic approaches use dependence relations of constitution to explain a system's macrolevel activity Ψ in terms of the behaviors of its constitutively relevant components, where the relation between the former and the latter is to be comprehended in terms of supervenience.[13] Simply put, all changes in Ψ are necessarily accompanied by at least one change in one of its component entities, whereas a change in a component entity is not necessarily associated with a corresponding change in Ψ.[14] Although the authors do not exclude that a tailored notion of causation might allow for interlevel causation in some sense, they argue that "there is no need to extend the word 'causation' to cover cases of this sort so long as one can describe the putative case of interlevel causation without remainder without doing so" (Craver and Bechtel 2007, 554).

To see why conceiving of interlevel relationships as causal might be dubious, consider the temperature of a gas, which is a measure of its kinetic energy content. The molecules in a hot gas move at high speeds and have high kinetic energy, whereas those molecules in a cold gas move at slow speeds and have low kinetic energy. The explanation that the cooling of the gas works by slowing down its component molecules is acceptable, so long as one refrains from maintaining that the slowed molecules stand in a causal relationship with the low temperature. A causal explanation would be mistaken: low temperature is constituted by the decreased kinetic energy of the molecules but is not causally produced by it.

Naturally, this carries implications for how an ideal intervention is conceived. Although the asymmetrical nature of causal relations means that it is only possible to unidirectionally intervene to change X and thereby change Y, but not vice versa, the symmetrical nature of a constitutive relation means that applying Woodward's unidirectional manipulability criterion will not be able to establish a constitutive relationship.[15] For this reason, Craver (2007) provides a sufficient condition for interlevel constitutive relevance. His mutual manipulability account (MM) of constitutive relevance is intended to characterize how scientists experimentally distinguish genuine components from causal factors. The parts and their activities are constituents of the phenomenon if the relevant interventions uncover mutual difference-making. As Craver (2007, 152–153) puts it, a part qualifies as a constitutively relevant component "when one can wiggle the behavior of the whole by wiggling the behavior of the component and one can wiggle the behavior of the component by wiggling the behavior as a whole. The two are related as part and whole and they are mutually manipulable." More precisely: "**MM** (i) instances of Φ are parts of S (ii) in the conditions relevant to the request for explanation there is some change to Φ that changes Ψ; and (iii) in the conditions relevant to the request for explanation there is some change to Ψ that changes Φ" (Craver 2007, 153). Condition (i) requires that X be a (spatiotemporal) part of the system, the behavior of which it explains, whereas conditions (ii) and (iii) are spelled out in terms of Woodward's (2003) notion of intervention. So Φ is constitutively relevant to Ψ if the relata stand in a part-whole relationship and if they are mutually manipulable, such that there is a possible ideal intervention on Ψ under which Φ is altered and vice versa. As expressed by conditions (ii) and (iii), constitutively relevant relations are established by executing interventions on a phenomenon in a top-down manner and on the parts in a bottom-up manner (figure 5.2). Bottom-up experiments

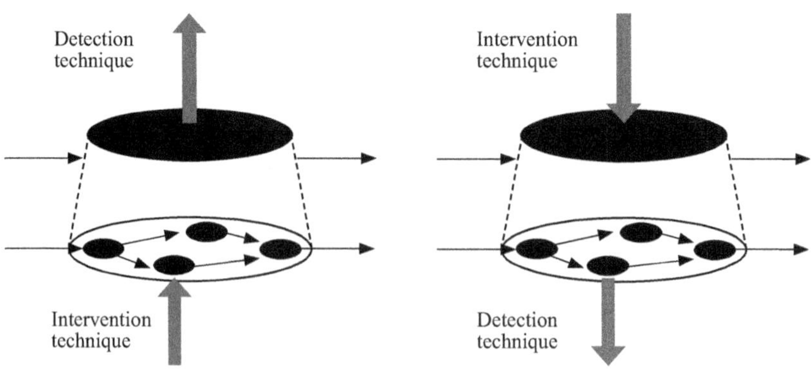

Figure 5.2
Interventions for testing constitutive relevance. Adapted from Craver 2007, 146, and Craver and Darden 2013, 126.

(captured by condition [ii]) intervene on putative lower-level components (often by boosting or reducing its activity) while tracking changes in the phenomenon at a higher-level. More precisely, experiments of this sort typically alter the value of the lower-level variable Φ while tracking changes in Ψ_{out} (output variable; Harinen 2014). In contrast, top-down experiments (captured by condition [iii]) intervene on the level of the overall phenomenon (often by engaging a subject in a task) while tracking changes on the lower-level of putative components. Experiments of this sort typically change the value of the higher-level variable Ψ_{in} while tracking changes in the value of the relevant lower-level variable. Although determining the adequate combination of bottom-up and top-down interventions will vary from case to case, demonstrating interlevel constitutive relevance necessitates performing both top-down and bottom-up experiments (Craver and Darden 2013).

As was the case with M, much turns on what qualifies as an ideal intervention. Analogous to Woodward's (2003) account, the basic idea is that an ideal intervention I on Φ is such that if the change in the value of Φ changes Ψ, it only does so via the change in Φ (Craver 2007, 154). This implies that intervention I

(I1c) does not change Ψ directly;

(I2c) does not change the value of some other variable Φ′ that changes the value of Ψ except via the change introduced into Φ;

(I3c) is not correlated with some other variable M that is causally independent of I and also a cause of Ψ; and

(I4c) fixes the value of Φ in such a way as to screen off the contribution of Φ's other causes to the value of Φ.

It is important that interventions satisfy these ideality conditions because otherwise it is not possible to isolate the variables. For instance, if I on Φ_1 also changes the value of Φ_2, while Φ_2 stands in an off-path causal relation to Ψ (i.e., not the path via which Φ_1 is causally related to Ψ), then no answer will be offered as to whether Φ_1 or Φ_2 are difference-makers with respect to Ψ.

In all, MM offers a sufficient condition for establishing constitutive relevance between the whole and one of its component parts. It adequately mirrors the experimental practices in the relevant scientific fields, but the appeal to manipulability to evaluate both intralevel causal and interlevel constitutive relationships in mechanisms is not unproblematic. Although intervention and its effects occur at different levels, interlevel constitutive relations are evaluated in the same way as intralevel causal ones. However, because constitution is understood as a noncausal interlevel dependence relation, Craver and Bechtel (2007, 547) add that interventions involve "mechanistically mediated effects" that are "hybrids of constitutive and causal relations in a mechanism, where the constitutive relations are interlevel, and the causal relations are exclusively intralevel." For example, what might seem like downward or top-down causation in fact comprises a relation of constitution between a purported higher-level cause and some lower-level part, which stands in a causal relationship to the lower-level effect. So in both bottom-up and top-down studies, what at first might look like interlevel causation can be understood in terms of intralevel causation combined with an interlevel constitutive relation. When interpreted in this manner, there is no need to worry about objections to top-down causation as plainly incoherent or as positing "spooky forces" exerted on components (Craver and Bechtel 2007, 548).

5.4 *Active Scaffolding (MM)*

In light of the demarcation problems investigated in chapter 4, MM looks like a promising candidate for specifying forms of ASC. MM has been developed independently and offers an experimental procedure that could help determine whether experimental results reveal constitutive or causal relations of dependence. In part for these reasons, some authors have argued that it could help demarcate the realm of the cognitive (Kaplan 2012; Zednik 2011; Kirchhoff 2017).

Advocates of EC have discussed studies by Dana Ballard and her group (Ballard, Hayhoe, and Pelz 1995) on the function of eye saccades in the performance of memory-demanding tasks (Clark 2008). Subjects were asked to complete a task (copying a pattern of blocks on the monitor by picking blocks from the "resource area" and arranging them in the "work area") while their eye movements were recorded. Although this top-down intervention engaged subjects in a memory task that prompts repeated saccades, the first finding was that subjects performed significantly more saccades (both to and from the provided pattern) than would be predicted if they had used a memory-intensive tactic that keeps the information about the provided pattern readily available while constructing the new pattern.[16] In the ensuing bottom-up intervention, freely saccading eye movement was prevented by instructing subjects to fix their gaze on the middle of the screen. Subjects in the fixed gaze condition required approximately three times longer to finish the assignment than did subjects in the unconstrained condition. Because different sizes of blocks were used across trials while the average completion times remained unchanged, the increased completion times were not due to limited visual access to the blocks in the constrained condition. The authors of the study concluded that the eye saccades are constituents of the cognitive process of block matching (Ballard, Hayhoe, and Pelz 1995, 71), which appears to be supported by MM. The top-down intervention is associated with changes in the saccades, whereas bottom-up intervention in changing the saccades is associated with changes in block matching. For this reason, some have argued that mutual manipulability is satisfied and that the saccades are constitutively relevant to block matching, establishing the existence of extended cognition based on empirical evidence (Kaplan 2012).

The initial attempt to specify ASC proceeds in similar ways, using MM as a guide. Thus, using mutual manipulability as expressed in three conditions as a guide, we could say that X is actively scaffolded onto Y if X and Y are mutually manipulable such that they fulfill conditions (i), (ii), and (iii) of MM. However, this enterprise immediately runs into a difficulty with condition (i), according to which "X is a part of S" (Craver 2007) or "contained within S" (Craver and Tabery 2017), which indicates that parthood is understood in terms of spatial inclusion. But the exact nature of the requirement that putative components be contained within S has remained underspecified and runs the risk of becoming too inclusive or exclusive. Two issues are worth noting here. First, (i) generates a problem for ASC. The requirement of parthood, a particular metaphysical relationship between the scaffold and the scaffolded, seems misplaced when the nature of that relationship is itself under debate. Assuming a parthood relationship would thus beg the question and appear problematic in light of the discussion in chapter 4.[17] To avoid these problems, it is best to proceed with what Wheeler calls a "locationally uncommitted" approach, which requires bracketing or changing (i). Second, (i) also leads to internal tensions in MM.[18] MM is open to the idea that mechanisms do not respect the boundaries of organisms, citing the work of some prominent proponents of extended cognition. Consider the following passage:

> Some mechanisms are entirely contained within physical compartments, such as a nucleus, or a cell membrane, or skin. ... However, mechanisms frequently transgress compartmental boundaries. The mechanism of the action potential relies crucially on the fact that some components of the mechanism are inside the membrane and some are outside. ... Likewise, many cognitive mechanisms draw upon resources outside of the brain and outside of the body to such an extent that it may not be fruitful to see the skin, or the surface of the CNS, as a useful boundary (as Haugeland 1995; Wilson 1995, 2004; Clark 1997; and Clark and Chalmers 1998 emphasize). (Craver 2007, 141)

The proposal that components of the mechanism can be external to the relevant system accords well the view that though mechanisms consist of objective structures, their boundaries are drawn in an explanandum-relative fashion. As Craver (2009, 590) maintains, "Stable spatial relations, I conclude, are neither necessary nor sufficient to define the boundaries of mechanisms." But if the mechanism can include parts beyond the boundaries of the organism, then the function of (i) is not clear.

To avoid this knotty issue, one could opt for a different, perhaps more relaxed account of parthood, or drop requirement (i) altogether. Speaking against the first solution, Leuridan (2012) finds that all plausible accounts of parthood generate problems for MM. This speaks in favor of the second solution, which lets mutual manipulability as expressed in conditions (ii) and (iii) serve as a criterion:

Active Scaffolding (MM): X is actively scaffolded onto Y if X and Y are mutually manipulable such that they fulfill conditions (ii) and (iii) of MM.

Dropping requirement (i) from the mutual manipulability account comes at a price. It leads to the result that circular or bidirectional causation can qualify as constitutive relevance (Leuridan 2012, 413). The question whether this is a genuine obstacle will be dealt with later; for now, let us attend to a problem that ASC imports from MM.

5.4.1 Surgical Interventions?

Many have noted that applying manipulations initially designed for analyzing causal relations to constitutive relations across levels is problematic in several ways (see Leuridan 2012; Harinen 2014; Romero 2015; Kästner 2017, ch. 6; Krickel 2017; Baumgartner and Gebharter 2016; Baumgartner and Casini 2017). One fundamental problem is that part-whole relationships conflict with the kind of "surgical" intervention contained in Craver's mutual manipulability criterion. Simply put, it is impossible to intervene on Ψ while leaving all Φs untouched (see figure 5.3a).

Instead, as Baumgartner and Gebharter (2016) argue, interventions will necessarily be "fat-handed" (figure 5.3b): they will inevitably affect multiple variables at the same time because any intervention on Ψ is necessarily an intervention on at least one component. As mechanisms supervene on their constituents, every cause of Ψ is necessarily a common cause of both Ψ and at least one component, but not vice versa, because changes in the supervenience basis are possible without changes in the supervening property.[19] To illustrate, suppose an intervention on the macrolevel variable Ψ with respect to Φ_1 shows a correlation between changes in Φ_1 and Ψ. The interventionist framework maintains that such a correlation indicates either a causal relation or a covert factor exercising causal influence on both. Thus, intervention I on Ψ could induce change in Φ_1 via causal pathway (a) or (b):

(a) $I \rightarrow \Psi \rightarrow \Phi_1$

(b) $\Psi \leftarrow I \rightarrow \Phi_1$

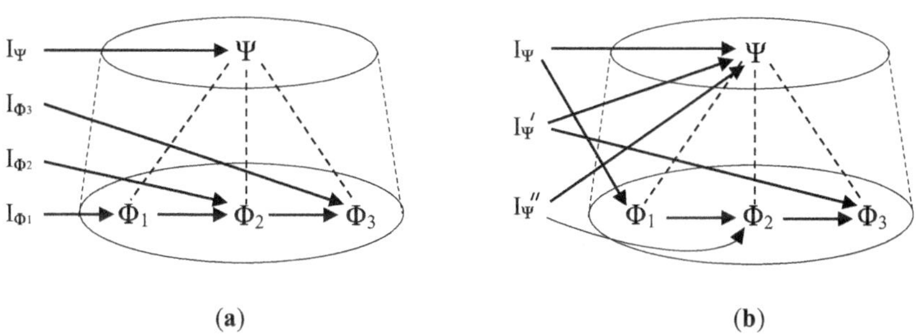

Figure 5.3
(a) Surgical interventions required by MM. (b) Possible fat-handed interventions. Adapted from Baumgartner and Gebharter 2016.

Given that constitution is understood as a noncausal relation, (a) is excluded. This leaves us with (b) and thus an explanation of the correlation in terms of a common cause, undermining the idea of ideal interventions into phenomena constituted by mechanisms. However, (b) may establish mutual manipulability via common cause, but it does not provide evidence of the existence of constitutive dependencies. The implications for ASC's appeal to MM are not difficult to see. Let Ψ represent the cognitive phenomenon of block matching, Φ_1 the underlying neural processes, and Φ_2 the saccading activity. The relevant possibilities are as follows: (1) Ψ is constituted by Φ_1 alone, (2) Ψ is constituted by Φ_1 and Φ_2, and (3) Ψ is neither constituted by Φ_1 nor Φ_2.

Offering support for the existence of extrasomatially scaffolded cognition would demand establishing (2) by way of performing the suitable surgical interventions. If intervention I on Ψ with respect to Φ_2 that holds off-path causes of Φ_2 fixed is associated with changes on Ψ with respect to Φ_2, then I is the common cause Ψ with respect to Φ_2. However, this intervention could induce change in Φ_2 via pathway (a) $I \rightarrow \Psi \rightarrow \Phi$ or (b) $\Psi \leftarrow I \rightarrow \Phi$. Although (a) is excluded by the assumption that constitutive relevance is noncausal, (b) offers a common-cause explanation that does not provide evidence of the existence of constitutive dependencies. Consequently, the studies *cannot establish constitutive relevance of the saccades*: the matching task does not qualify as a top-down surgical intervention on Ψ, and gaze fixation does not qualify as a bottom-up surgical intervention on Φ_2. The studies thus fail to determine that (2) is correct. Worse, they also fail to establish (1). Because the possibility of surgical interventions is necessary for constitution, and because ideal surgical top-down interventions are not possible, we are forced to accept the unattractive option (3), according to which cognitive phenomena are not constituted by neural activity either.

Overall, though MM can generate unconfounded data, it appeals to surgical interventions that cannot be performed and fails to identify the constituents of cognitive processes. This has repercussions for the aim to understand ASC in terms of constitutive relevance established by the existence of appropriate mutual manipulations. To avoid having to accept (3), and to clarify ASC, some of the requirements of MM need to be revised. The two most obvious involve rethinking the relationship between causation and constitution or relaxing the requirement that interventions need to be surgical. Starting with the latter option, the following sections will explore these possibilities.

5.4.2 *Active Scaffolding (MM*)*

We may accept that interventions are necessarily fat-handed and modify or abandon the surgicality requirement for interventions. Woodward's adjusted characterization of ideal interventions can be deployed for this task, which relaxes the requirements for an intervention to qualify as ideal while allowing exemptions in case supervenience relations hold. Woodward (2015) drops a constraint and allows for interventions to have several target variables, though only if they stand in a supervenience relation. The adjusted view is that "an intervention on a variable X also should be treated as automatically changing (indeed as also an intervention on) the supervenience base SB(X) of X, with SB(X) changing in whatever way is required by the supervenience relation between X and SB(X) (or as changing in some way or other that is consistent with the supervenience relation if there are multiple possibilities for the state of SB(X) under the change in X)" (Woodward 2015, 333). Accordingly, an ideal intervention on X does not affect any off-path variables—except of course for variables that X constitutively depends on. Consequently, when evaluating whether X causes Y, the variables kept fixed should not comprise X's supervenience base and the supervenience base for Y (Woodward 2015, 333–334). Translated to the notation used in the mechanist literature, I on Ψ is a permissible, fat-handed intervention with respect to Φ, even if I causes Φ along a path that either goes through Ψ or through variables that stand in a supervenience relationship to Ψ.

With these changes, ideal interventions are possible as fat-handed interventions, which serve as common causes of non–causally related variables. We can now use MM* to specify ASC, replacing MM. This solves a crucial problem by no longer appealing to surgical interventions that cannot be performed, and it looks promising for the aim to understand active scaffolding relationships in terms of constitutive relevance established by mutual manipulability. However, the revision comes at a relatively steep price. In MM*, the data provided by a fat-handed intervention on target variables is relatively unhelpful to identify the relationship between these variables. Proceeding in this manner risks dissolving the decisive difference between constitutive and causal relevance (Baumgartner and Gebharter 2016, 747).

This is because in case an intervention I generates interlevel correlations, a suitable explanation of the arising correlation can be given by invoking the fact that I is fat-handed and affects both levels. Even if interventions I_Φ and I_Ψ lead to changes in Ψ and Φ, no inference is warranted to Φ being constitutively relevant for Ψ. A system featuring constitutive interlevel dependencies and one featuring causal interlevel dependencies will involve indistinguishable correlations under I_Ψ and I_Φ. Data generated under I_Ψ and I_Φ will be explicable by a model positing a constitutive relevance between Ψ and Φ, but also by a model positing a causal relevance between Ψ and Φ.

In all, MM* successfully loosens the constraints on interventions to accommodate that manipulations producing changes on both levels are necessarily fat-handed. But as a result, the detection of interlevel mutual manipulability in MM* becomes relatively uninformative because it is explainable by the mere fact that manipulations are fat-handed. There is, however, still a way in which interlevel mutual manipulability is informative. Quite simply, unlike in the case of MM, constitutive relevance on MM* is indicated not by the possibility of surgical interventions but by their impossibility. Appropriate surgical interventions could break causal dependencies between Ψ and Φ, but no interventions could disrupt constitutive dependencies between Ψ and Φ because they are *unbreakably coupled* via common causes (Baumgartner and Casini 2017; Baumgartner and Wilutzky 2017).

Having clarified how MM* could relax the requirement that interventions need to be surgical while still retaining the difference between causal and constitutive relations of dependence, we may now return to the overall goal to specify ASC. At first, the view that the identifying feature of a constitutional relation is being unbreakably coupled via common causes (and thus persisting across all variable set expansions) appears productive for the enterprise, as it retains a close connection to explanatory practices and carries empirical implications. Moreover, it emphasizes that causation and constitution are both difference-making relationships. This gives us reasons to think that MM* is useful to specify and test for ASC, without having to rely on the kind of intuitions that were explored in chapter 4.

5.4.3 Two Challenges

Using MM led to the realization that the surgicality requirement had to be relaxed, but the implementation of this change in MM* leads to two challenges that both are closely related to the question of constitutive or causal dependence.

5.4.3.1 A Difference That Makes a Difference? The first challenge is that MM* does not offer a way to conclusively attest constitutive relations, but only abductive inference to constitutive dependence via many bottom-up and top-down experiments, because underdetermination problems cannot be fully defused. In MM, the existence of a single pair of bottom-up and top-down experiments warrants inferring constitutive relevance, but MM* maintains that constitutive relations can only be inductively corroborated, not conclusively verified. Although several bottom-up and top-down experiments failing to break coupling (via common causes) warrants abductive inference to constitutive relevance, no amount of experiments would be able to rule out that further experimentation might reveal surgical causes. In this respect, establishing a constitutive relation between Φ and Ψ is like establishing causal irrelevance between them: both can only be inductively corroborated. Granted, compared to a causal analysis, constitution offers a superior explanation for correlated changes in Φ and Ψ because constitution also explains why interlevel interventions fail to break the dependencies between them. Satisfying MM* by offering abductive evidence for constitutive relevance is in this sense productive. However, the question is how much of a difference this makes from a pragmatic perspective. In MM*, inference to constitutive relations "requires an extended test series exploring the whole space of possible ways of breaking the coupling of macro and micro levels" (Baumgartner and Casini 2017, 217). Overall, though MM* is still useful to specify ASC, the fact that no indication is available about the amount of interlevel experiments that would warrant abductive inference to constitutive relevance detracts from its productivity.

The overall framework that this chapter operates with offers possibilities to deal with these challenges in a satisfactory manner. As to the first challenge, we may say that while there clearly is a genuine theoretical issue here that is worth addressing, it is relatively negligible from the point of view of the overall goals of this chapter. Bluntly put, whether the relevant interlevel relation is mutually manipulable because the relation is unbreakably coupled or bidirectionally causal is relatively inconsequential from a pragmatic standpoint.[20] A robust bidirectional causal relationship may in this view thus qualify as quasi-constitutive by courtesy, in virtue of its resemblance to constitution.

5.4.3.2 Interlevel Causation The second challenge is that though MM* renders ideal interventions possible, it also allows relations of interlevel manipulability to qualify as causal. To see how, recall that when introducing the idea of mechanistic levels, we said that Ψ can be comprehended as a multifaceted input-output relationship characterized by some regularity between the values of variables Ψ_{in} and Ψ_{out}. We perform a top-down intervention by altering the value of Ψ_{in} and identify alterations in the value of Φ_3. Correspondingly, our bottom-up intervention on Φ_3 changes its value and we identify changes in the value of Ψ_{out}. In MM, I on Ψ_{in} with respect to Φ_3 is not ideal. Either Φ_3 is in the supervenience base of Ψ_{in} (which would mean that I changes it directly), or it is causally related to at least one of the variables in the supervenience base of Ψ_{in} that I changes. In MM*, I on Ψ_{in} with respect to Φ_3 can be ideal, if I on Ψ_{in} does not directly change Φ_3 (because Φ_3 is in the supervenience base of Ψ_{in}) but only some variables in the supervenience base of Ψ_{in} that stand in a causal relation to Φ_3.[21] This, however, opens up the possibility that relations of interlevel manipulability can be causal. If the top-down intervention on Ψ_{in} with respect to Φ_3 and bottom-up I on Φ_3 with respect to Ψ_{out} are ideal in the sense of MM*, then the relationship between Ψ_{in} and Φ_3 counts as causal, not constitutive.

Along somewhat similar lines, Krickel (2018) presents an adjusted view of mechanistic levels that renders possible interlevel causation. To stress the temporal extendedness of activities, Krickel (2018; Kaiser and Krickel 2016) introduces the notion of entity-involving occurrents (EIO), which can be split into spatial and temporal EIO parts. Crucially, she maintains that spatial EIO parts are situated at a lower level than temporal EIO parts. Because they are located on different levels, they are not related as parts and wholes, which makes an interlevel causal relation between them possible. As Krickel (2018, 15) puts it, "If we take mechanistic interlevel causation to consist in causal interactions between the components of the mechanism and temporal EIO-parts of the phenomenon, we can make sense of interlevel causation in mechanisms."[22]

Although these accounts illustrate how interlevel causation may be possible, the approach in the following is a different one. The point is that interlevel causation is less threatening than it may first seem. Presenting a comprehensive set of arguments is beyond the scope of this chapter, but in what follows, some reasons will be offered for thinking that the differences *between constitution and bidirectional causation* may not be as firm as Craver and Bechtel seem to think. To see why, one could start by pointing out that causation and constitution are both difference-making relationships and that both causal and constitutive explanations trace networks of counterfactual dependence (Ylikoski 2013). Moreover, constitutive explanations account for causal capacities and the manipulation involved in experimental testing of constitutive relations is causal. To dig deeper, it is helpful to review the motivation for maintaining that causal and constitutive relations are conceptually different and mutually exclusive relations. Craver and Bechtel (2007) point out major differences between casual and constitutive relationships:

(i) *Distinctness:* Constitutive relevance holds among wholes and their parts, whereas causally dependent events X and Y are distinct from each other.

(ii) *Temporal overlap:* Changes in the marco- and microbehaviors of a mechanism temporally overlap, whereas causes temporally precede their effects.

In light of these differences, the point is that comprehending interlevel relations between Ψ and Φ would violate dominant ideas connected to the concept of causation (Craver and Bechtel 2007, 554). For instance, it would be wrongheaded to maintain that the turning of the wheels of a car causes the car's driving behavior. Relative to the phenomenon, the behaviors of the wheels and the car are not distinct and their activities temporally overlap. But in many other cases, things are more complicated.

For example, researchers find mutual manipulability between depressed mood and muscle activation in a particular area of the gebellar region associated with expressing sadness. The findings seem to indicate mutual manipulability, yet it is not clear that undergoing depressed sadness (Ψ) and specific muscle activation in the gebellar region (Φ) is distinct in the sense required by (i). They appear to be on different levels, but in the literature on emotions, the dominant view does not regard muscle activation as a constituent of moods and emotions, so using the notion of parthood deployed in the mechanistic literature would not be applicable. The spatiotemporal region that the muscle activation occupies is contained in the spatiotemporal region that the emotion occupies, but the case does not license concluding that Φ is a spatiotemporal constitutive part of Ψ such that they cannot be causally related. Probing (ii) also fails to yield a straightforward answer. When performing bottom-up and top-down interventions in these sorts of cases, there will be a temporal delay between the changes in the macro- and microlevels. Whether or not the delay warrants denying temporal overlap would depend on further, independently justified qualifications.

These issues raise some questions as to whether it is productive to insist on a firm difference between stable bidirectional causation and constitution. These considerations offer at least preliminary reasons for adopting a view that softens the difference between bidirectionally causal and constitutive relevance. Of course, this move will be resisted by those who, like Craver and Bechtel (2007), want to describe putative cases of top-down causation as mechanistically mediated effects. However, though this solution does not require positing additional interlevel causal processes, it does not sidestep a version of the exclusion argument raised against top-down causation (Bechtel 2017; for a related criticism, see Eronen 2013).[23] Instead of comprehending the mechanism as a higher-level entity, one might object that it is better to speak of a collection of lower-level entities that are responsible for the entire range of causal activity, rendering higher levels epiphenomenal. Furthermore, the opponent could also wonder why one would assign the effect of, for instance, Φ_1 and Φ_2 to Ψ, when what Φ_1 and Φ_2 influence is Φ_3 and Φ_4, which, together with Φ_1 and Φ_2, jointly constitute Ψ.

So far, the argument was mainly meant to soften the difference between causal bidirectionally and constitutive relevance, to show that it may not be as strict as Craver and Bechtel hold. To make the case stronger that ASC does not import a problematic notion of top-down causation from MM*, we may further clarify why top-down causation in not necessarily problematic or mysterious.

One possibility is to stress that partial constitution (along with proper parthood) can be placed somewhere between identity and distinctness. Because distinct things cannot stand in a constitution relation, X being a constitutive part of Y means that X is neither identical with Y nor distinct from Y. The point is that this opens room for conceiving the relata in at least some relations of constitution as "causally entangled" (Zangwill

2012). This option revolves around the existence of an intermediate category of difference-making relation that transmits causal properties. Others have construed constitution as a species of causation. The idea is that if manipulability defines causal relationships, then constitutive relationships can be understood in terms of a bidirectional causal dependence. From this background, some suggest dropping the label *constitutive* and distinguishing instead between two forms of causal relevance: a strict intralevel relevance and an interlevel relevance (Leuridan 2012, 424).

There are also less radical alternatives available, which accept some minimalist notion of top-down causation in mechanisms but explicate it in terms of partial "constraints" (Kistler 2010; Bechtel 2017). Ψ can exert a downward constraint in the sense of limiting the space of possible values that lower-level variables can take on. Although Craver and Bechtel analyze the "enlisting" of parts by the whole system in terms of constitution, enlisting also lends itself to a causal interpretation, particularly if the temporally extended nature of the phenomenon is stressed. If Ψ is temporally extended, then the constraint on the lower-level variable via enlisting can be understood as transferring causal powers in a delayed manner,[24] without violating the principle that causes and effects are spatially and temporally distinct.

In fact, recent work by mechanists indicates willingness to allow for a top-down constraint construed along similar lines. Using the circadian mechanism as an example, Bechtel (2017) has recently deployed graphical representations to clarify how top-down constraints could look without rendering higher levels epiphenomenal. Bechtel argues that the circadian mechanism constrains various physiological processes like gene expression, thereby restricting the behavior of its parts and changing how they are affected by inputs. In short, "being situated in the mechanism constrains the behavior of the part" (Bechtel 2017, 271).

5.5 Conclusion

The notion of scaffolding offers a productive conceptual flexibility that can incorporate synchronic and diachronic forms of scaffolding. However, it also characterizes too wide a variety of relationships of dependence between cognition and the body or the environment. Even if the notion of active scaffolding excludes trivial forms of dependence, it is still consistent with distinct metaphysical relations. For this reason, this chapter explored relations of dependence by engaging influential accounts of causal and constitutive relevance. Instead of seeking guidance from the kind of approaches discussed in chapter 4, the goal was to offer a more neutral base and to retain proximity to explanatory practices in the relevant scientific fields. The chapter started by drawing on M, a manipulability (or interventionist) account of causal relevance. Because active scaffolding understood along the lines of M allows certain relationships to qualify as causal that do not display characteristic features of exemplary scaffolding relationships, further conditions of stability and specificity were added. The addition of these auxiliary conditions led to a manipulability account of active scaffolding, or *Active Scaffolding (M)*.

Some active scaffolding relationships are more complex, and their dependence is often mutual, with bidirectional difference-making that indicates a metaphysically distinct constitutive relevance. To capture this form of active scaffolding, this chapter drew on the literature on mechanistic approaches to explanation and Craver's (2007) account of mutual manipulability (MM) to offer a mutual manipulability account of active scaffolding, or *Active Scaffolding (MM)*. This move proved productive, but ASC imported some difficulties that can be traced back to the fact that MM appeals to the notion of intervention in M to evaluate both intralevel and interlevel constitutive relationships. MM* could relax the requirement that interventions need to be surgical, but it only licenses abductive inference to constitutive dependence via extensive bottom-up and top-down testing, while it also allows relations of interlevel manipulability to qualify as causal. These challenges led to revising active scaffolding in a way that tempered the difference between bidirectionally causal and constitutive relevance.

6 Complex Scaffolding

6.1 Bottom-Up and Top-Down

Previous chapters have highlighted evidence that cognition can be scaffolded in several ways, offering supporting structures that cognitive processes can utilize. Having clarified the relationship of dependence in active scaffolding relations, this chapter is dedicated to an in-depth investigation of *complex scaffolding*, which highlights how the acquisition, structuring, and meaning of certain concepts is scaffolded onto the sensorimotor system. The goal is to generate insight into cognition, as well as potential implications for understanding and treating mental disorders. More precisely, we are interested in how cognitive activity X being actively scaffolded onto a structure or process Y allows that manipulating Y can affect X, and perhaps vice versa. If both the scaffolding thesis and the phylogenetic story about the massive reuse of existing neural circuits are accurate, then we may anticipate *Active Scaffolding (MM)*, with interesting uni- and/or bidirectional cross-domain effects between sensorimotor and cognitive processes.

For this, the next sections offer a brief but systematic overview of relevant behavioral and neuroscientific research. Most of the experiments that will be surveyed use priming methodologies that aim to investigate bottom-up and top-down cross-domain effects. Consider, for instance, a term like *power*, which is habitually expressed in terms of space and spatial experience. To investigate the pertinent effects, there are two obvious possibilities, using what mechanist approaches would call *bottom-up* and *top-down interventions*:

Bottom-up: These studies typically expose participants to particular perceptual stimuli and study the extent to which this exposure has *bottom-up effects* on higher-order judgments. For example, a study has shown that the vertical position of group labels on a screen can influence judgments about which group possesses power, as well as the speed and accuracy of judgment (Schubert 2005). Such studies perform an intervention on low-level variables and can indicate the existence of unidirectional effects that flow in a bottom-up direction.

Top-down: These studies typically activate certain concepts or goals in participants and study the extent to which this has *top-down effects* on perceptual and sensorimotor experiences. For example, a study demonstrated that beliefs about power relations can modulate how participants experience vertical location (Giessner and Schubert 2007). Such studies perform an intervention on high-level variables and can indicate the existence of a unidirectional effect that flows in a top-down direction.

Importantly, though some effects are unidirectional (Casasanto and Boroditsky 2008), for the most part studies offer evidence for bidirectionality (Körner, Topolinski, and Strack 2015).

6.2 Some Evidence for Uni- and Bidirectional Effects

The following sections provide a summary of some of the most relevant research in this area, focusing on effects related to physical temperature, cleanliness, weight, texture, and hardness.

6.2.1 Hot-Headed and Cold-Hearted

It is common to deploy temperature concepts when characterizing other persons, social relationships, and interactions. Moreover, when individuals form first impressions and make fast social judgments about trustworthiness, they proceed along a few dimensions, among which interpersonal warmth/coldness is the most important (Fiske, Cuddy, and Glick 2007). *Warmth* in this context should be understood as referring to a cluster of traits including friendliness and helpfulness, whereas *coldness* refers to opposite traits, and coldness-related expressions (like "give the cold shoulder") describe social rejection or exclusion.

In two studies in which the experimenter was blind to the hypotheses of the respective studies, Williams and Bargh (2008a) have shown that tactile experiences of physical warmth trigger the activation of the concept of interpersonal warmth, which then exerts an influence on judgments and behavior. Participants were requested to hold a hot or a cold cup of coffee for a short time. Although this did not influence judgments about traits unrelated to warmth and coldness, participants holding the hot beverage rated the target person as substantially warmer. The second study extended the reach from interpersonal judgment to behavior and showed that such priming also affects the participant's behavior toward target individuals. Participants primed in the "cold" condition tended to choose a reward for themselves, while participants in the "warm" condition preferred the option of rewarding a friend.

While this study shows a unidirectional, bottom-up effect in which experiences of temperature affect judgments of trustworthiness and prosocial behavior, other studies demonstrate top-down effects. Zhong and Leonardelli (2008) show that participants recollecting an experience of social exclusion estimate the room temperature to be lower than participants recollecting an experience of social inclusion, leading the authors to conclude that being rejected induces the perception of coldness (Zhong and Leonardelli 2008, 840). Also, participants experiencing a form of social exclusion (induced via online interaction) conveyed more desire for warm beverages and food than participants experiencing social inclusion. Correspondingly, being socially excluded in a simple online ball-tossing game results in a decrease in skin temperature, whereas physical warmth (holding a cup of warm tea) helps alleviate the adverse feelings linked to social exclusion (IJzerman et al. 2012).[1]

In addition, an investigation of heat metaphors (e.g., being hot-headed) found that visual representations of heat ease the employment of conceptual knowledge related to anger, whereas priming thoughts related to anger leads participants to overestimate the actual room temperature (Wilkowski et al. 2009). A related study by Gockel, Kolb, and Werth (2014) on the relationship between interpersonal perceptions of coldness and warmth examined the influence of ambient temperature on judgments of criminals (related to both attributing criminal intent and inferring criminality). Exposed to different room temperatures within a comfort zone (corresponding to upper, medium, or lower levels), participants were invited to inspect pictures of individuals who had committed a crime. Participants exposed to the high temperature were more likely to ascribe impulsive crimes than participants exposed to the low temperature (consistent with an assessment of them as more hot-headed), and the latter group was more likely to ascribe premeditated crimes to the criminals, as well as crimes resulting in a more severe penalty (consistent with an assessment of them as more cold-hearted). Although ambient temperature clearly influenced the judgment of criminals, changes in affective states could not explain the results. The results are further corroborated in a study exploring the influence of physical coldness on judging moral dilemmas (Nakamura et al. 2014). Participants exposed to cold room temperature were more likely to opt for a utilitarian judgment—not because coldness promoted more cool-headed reflection, but because it helped reduce levels of empathetic concern, rendering participants more cold-hearted in their judgment of particular moral dilemmas.

Supporting these findings, some neuroscience studies indicate that the connection between social and physical warmth and coldness might have a solid anatomical basis in human beings, with the insular cortex implicated in processing physical temperature and psychosocial input about warmth associated with personality (Meyer-Lindenberg 2008; see Bargh and Shalev 2012 and Kang et al. 2010 for an fMRI study).

6.2.2 Dirty Hands and Clean Conscience

Olfactory experience may under certain circumstances influence judgment and behavior (Schnall, Benton, and Harvey 2008). When exposed to cleaning scents, participants not only are more proficient in accessing cleaning-related constructs, but also tend to maintain a cleaner environment (Holland, Hendriks, and Aarts 2005). Combining such results with the fact that *clean* and *pure* designate both physical and moral states and that morality is often expressed in terms of cleanliness and purity (metaphors like dirty hands, clean records, clean conscience, etc.), it may be expected that the strong association between moral and bodily purity would result in interesting effects. What might further fuel this suspicion is the way in which the act of physical cleansing has occupied a central place in religious ceremonies for thousands of years. For instance, baptism in Christianity links the symbolic cleansing of the body to a cleansing of conscience, consistent with the idea that being morally virtuous is to be physically clean and free from the filth of sin. Moreover, neurophysiological evidence indicates that contamination and moral transgression trigger the same face muscles and corresponding brain networks (Chapman et al. 2009; Cannon, Schnall, and White 2010; Borg, Lieberman, and Kiehl 2008; Lee et al. 2015).

Zhong and Liljenquist (2006) demonstrate that recalling a memory of unethical behavior (or reading a short story describing an unethical act) augments the accessibility of concepts related to cleansing and that the menace to moral purity generates in participants a desire for physical cleansing, which manifests through preferences for cleansing products. When given a choice between a pencil and an antiseptic wipe after the recall task, participants who recollected an instance of unethical behavior were more likely to choose the antiseptic wipe than those who recollected ethically correct behavior. Although exposure to moral transgressions motivates a desire for physical cleansing, the study has also found that physical cleansing actually can work as a substitute for moral purification: physical cleansing compensates for moral impurity and effectively assists in restoring a sense of moral integrity. When participants physically cleansed after recalling an unethical act, they reported reduced moral emotions, and their direct compensatory behavior for unethical acts dropped by almost 50 percent, indicating that physical cleansing actually can "wash away" the feeling associated with moral wrongdoing. This is further supported by the finding that individuals make larger donations to charity prior to, rather than after, bathing for religious purification (Lobel et al. 2014), as well as by a study exploring the influence of physical cleansing on a physiological level by examining pupil size (Kaspar, Krapp, and König 2015).[2]

Other studies provide additional support. For instance, acting violently in a virtual environment when playing video games caused participants to select hygiene-related products instead of non-hygiene-related products, perhaps because the video game violence was, on some level, threatening to the moral integrity of the participants (Gollwitzer and Melzer 2012). The effects may be stronger in individuals with obsessive-compulsive disorder (Reuven, Liberman, and Dar 2014), but they are also mode-specific: after performing an immoral act using their hands, participants preferred hand sanitizer over other products, whereas after performing an immoral action using their mouth (telling a lie), participants preferred mouthwash (Lee and Schwarz 2010).[3]

For some neurobiological support, consider an fMRI experiment by Denke et al. (2014). Participants first read and enacted scenarios that involved either lying or telling the truth and then rated the desirableness of different products. Participants who had performed an immoral act rated cleansing products as more appealing compared to those who had enacted a moral deed. Both during the rating of cleansing products (but not products unrelated to cleansing) and after the enactment of the unethical behavior, the effect was accompanied by activation in sensorimotor brain regions. The authors conclude that moral-purity metaphor is tightly linked to certain sensorimotor activations and suggest that sensorimotor grounding is intrinsic to moral cognition (Denke et al. 2014, 6). Such a conclusion is in accord with earlier work on the neural foundations of metaphors (Gallese 2005; Gallese and Lakoff 2005).

If this is true, then one might expect that the sense of cleanliness would have an impact on the perceived weightiness of moral wrongdoings committed by others. This is, in fact, what a study by Schnall, Benton, and Harvey (2008) revealed: in a scrambled-sentences task, participants primed with concepts of cleanliness made less severe moral judgments compared to controls primed with neutral words. It has previously been established that feeling disgust tends to bring about more severe moral judgments, and that disgust is reduced if participants physically cleanse themselves after the experience (Wheatley and Haidt 2005), but this study also showed that physically cleansing after experiencing disgust leads to less severe moral judgments. One difficulty with this study is that the priming occurred in a way that left open whether the target or the self was implicated. However, a more recent study confirmed that a sense of cleanliness affects and improves one's moral self-assessment, which, in turn, results in stricter judgments on matters that have potential moral implications (Zhong, Strejcek, and Sivanathan 2010).

In addition to affecting moral judgment, the perception of physical cleanliness also appears to motivate virtuous behavior. In an anonymous trust game, participants had a choice between exploiting the sender by keeping an entire sum and honoring the trust by returning some of the money. Compared to those in a baseline condition, participants placed in a clean-scented condition returned considerably larger amounts of money and chose to reciprocate trusting behavior. Also, compared to controls, participants in the clean-scented condition articulated stronger motivation to volunteer for a charitable nonprofit organization and to donate money (Liljenquist, Zhong, and Galinsky 2010). These results are in accordance with Lee and Schwarz's (2012) study that revealed bidirectional effects related to the typical metaphoric expression of social suspicion: "smells fishy." Exposure to fishy smells stimulated suspicion in participants and undercut their willingness to engage in cooperation and in trust-based economic exchanges. Also, participants who engaged in a two-investor public goods game and were exposed to fish odors were less inclined to trust others, to act prosocially, and to contribute to shared resources. These studies show a bottom-up influence running from sensory perception to cognition and behavior, but socially induced suspicion also can cause top-down, metaphor-specific effects, increasing participants' accuracy in labeling fishy smells.

Overall, these studies provide evidence that simpler classifications based on bodily experience (clean vs. dirty) scaffold complicated social categorization and judgment (moral vs. immoral) and that the activation of related embodied concepts exhibits bidirectional effects. Experiences of physical cleanliness influence moral judgment and behavior, whereas moral cognition exerts an influence on the body.

6.2.3 The Weight of Guilt

The haptic dimensions of weight, texture, and hardness are also associated with more complex concepts via metaphors. For instance, in everyday language there is a clear metaphorical association between the experience of weight in terms of heaviness or lightness, and concepts of seriousness and importance. We often talk about "heavy responsibilities," "weighing the importance of options," "matters that carry weight," and the "gravity of the situation." There is also a link between experiences of texture and certain higher-order judgments that transpires in expressions like "having a rough day" or "using coarse language," whereas the connection between the tactile experience of a hard surface and the social experience of someone's inflexibility is highlighted in descriptions of persons as "hard-hearted" or "a rock." From this background, we might expect that conceptualizations of more complex moral and sociopolitical judgments might be scaffolded onto physical sensations of weight, texture, and hardness, and hence that basic tactile sensations have at least a unidirectional, metaphor-specific effect on higher-order, social cognitive processing.

First, behavioral experiments provide support for the thesis that "importance" is firmly grounded in physical experiences of heaviness or weight. A study by Jostmann, Lakens, and Schubert (2009) hypothesized that experiences of weight exert influence on the choice of issues considered important by the participants. Participants with a heavy clipboard considered self-determination ("having a voice") as more central, and they were more deeply engaged in cognitive elaboration than participants with light clipboards. Moreover, they perceived the topics that they were asked to rate as more important (Jostmann, Lakens, and Schubert 2009, 1173). A study by Ackermann, Nocera, and Bargh (2010) supports these results: when participants assessed a job applicant by evaluating a resume on either light or heavy clipboards, consistent with the "heavy is serious" metaphor, the participants with heavy clipboards rated the applicant as better overall, more important, and more seriously interested in the position. They also exhibited a higher rate of preference for providing funding for solutions to critical problems (e.g., air pollution standards). A more recent study shows that cognitive performance is reduced and the perceived effort is increased when participants hold a heavy clipboard (Kaspar and Vennekötter 2015).

Correspondingly, studies focusing on the "weight of guilt" metaphor uncover strong links between moral transgressions and sensations of weight. Participants recalling a memory of an immoral act reported higher subjective body weight than the control group, the members of which either did not recall a memory or recalled a memory of a moral act. The bodily sensation of physically being weighed down is mediated by heightened feelings of guilt (but not other negatively valenced emotions) and has top-down effects on judgments related to physical behaviors, leading to an overestimation of the effort required to accomplish physical tasks (Day and Bobocel 2013). Kouchaki, Gino, and Jami (2014) demonstrated that experiences of heaviness can intensify the experience of guilt. Participants carrying a heavy backpack reported feeling more guilt compared to participants carrying a light backpack, and they were also more likely to cheat less and to choose healthy snacks. However, carrying the heavy backpack did not affect the recollection of experiences that induce guilt, nor did it cause differences in terms of cognitive performance.

Although most of these studies find unidirectional, bottom-up effects, there is at least some indication that top-down effects also occur and that it is possible to reverse the direction of activation. In two studies by Schneider et al. (2011), participants were asked to hold and estimate the weight of a book, while a subset of participants were informed that the book was important. Participants in this subset estimated the book to be heavier (but not more expensive), showing that importance-related information about an object has an impact on judgments about its heaviness. Zhang and Li (2012) show that carrying a heavy object enhances the accessibility of the concepts "heavy" and "weighted," which results in an increase in judgments about importance. However, the authors stress that thinking about something as important does not necessarily need to affect the bodily experience of carrying weight and that only the effects between carrying weight and stress are properly bidirectional.[4]

Second, some studies explore the link between tactile sensations of roughness and the metaphorically connected concepts of "difficulty" and "harshness." One group of participants was asked to complete a puzzle consisting of pieces covered in rough sandpaper, while the other group was asked to complete the same puzzle with smooth (uncovered) pieces. When they subsequently read a description of an ambiguously valenced social interaction, those who had finished the rough puzzle assessed the interaction as more difficult than those who had finished the smooth puzzle. Moreover, when playing an "ultimatum game" after completing the puzzle, the bodily experience of roughness tended to foster compensatory bargaining (Ackerman, Nocera, and Bargh 2010). Such findings are supported by an fMRI study that tested whether certain higher-order processing recruits domain-specific sensory cortex activity (Lacey, Stilla, and Sathian 2012). The authors limited the investigation to metaphors related to texture and demonstrated that the texture-selective somatosensory cortex is activated when participants process sentences with textural metaphors.

Third, studies examining the link between haptic experiences of hardness and the metaphorically associated concepts of rigidity, stability, and strictness yield supporting results. In a study by Ackerman, Nocera, and Bargh (2010), participants were first asked to hold either something soft (blanket) or something hard (wood) and subsequently to examine an ambiguous boss-employee interaction to evaluate the employee's personality. Participants in the "hard" condition judged the employee to be stricter or more rigid than participants asked to hold the soft blanket, though without judging the employee more positively overall. In a second study, participants were passively primed by sitting on either a soft-cushioned or a hard chair. Participants assigned to hard chairs rated the employee as less emotional and more stable than did participants in soft chairs. When asked to complete a negotiation task that involved placing two bids on a car, those exposed to the hard chair were more rigid and less inclined to change their offers.

In all, consistent with the basic idea of complex scaffolding, embodied metaphors firmly associate the haptic dimensions of weight, texture, and hardness with complex concepts so that simple tactile sensations have a metaphor-specific effect on social cognitive processing and vice versa. Although the studies have provided evidence for uni- and bidirectional effects, there has been little focus on the role that individual differences have on these effects. Some recent evidence nevertheless suggests that interoception might modulate the effect of the haptic experiences on cognition so that individuals more sensitive to their bodily sensations display a stronger effect (Häfner 2013).

6.3 Application and Calibration

Complex scaffolding offers a phylogenetic and developmental account of how conceptual structure, the sensorimotor system, and cognition interact. Uncovering *Active Scaffolding (MM)*, the results showing uni- and/or bidirectional cross-domain effects (between sensorimotor and cognitive processes) open new possibilities for rethinking how embodiment may be relevant for understanding mental disorders. But whereas a systematic analysis of the empirical material helps comprehend the reach and practicality of the framework, it also renders palpable some challenges that help calibrate it.[5] The idea of complex scaffolding builds on the work of Lakoff and Johnson, but offering the best fit for the empirical material and for exploring mental disorders requires a number of changes.

The original work of Lakoff and Johnson, published prior to much of the empirical material discussed here, posits an asymmetrical bottom-up and top-down relation (Lakoff and Johnson 1999, 56; Lakoff 2008, 2012), which is consistent with the anti-intellectualistic account of cognition they envisage. Their account correctly predicts bottom-up influences from the sensorimotor domain to the cognitive domain but cannot accommodate top-down effects on the sensorimotor domain. In contrast, complex scaffolding within the framework of ASC harbors no commitment to positing an asymmetrical relation. This means that it faces no obstacles in explaining mutual manipulability.

Complex scaffolding also is a more limited account in several senses. It only applies to a limited number of metaphorical constructs and does not unearth universal mechanisms that characterize cognition in all human agents. There are a vast number of commonly used metaphors in natural languages; some of them are acquired by understanding complex semantic structures, whereas others take on a different (literary) meaning over time. It would be unwarranted to suppose that they all exhibit cross-domain mappings and give rise to bidirectional effects. To steer clear of potential pitfalls, it is beneficial to limit the scope of the account to commonly used metaphors that display a direct link to the bodily domain.[6] This means excluding many metaphors, like "a blanket of snow" or "a heart of gold," from consideration. Moreover, complex scaffolding accepts some degree of relativity to cultural background. One could initially assume that metaphors such as "morality is purity" would exhibit identical effects across cultural borders, but there are indications that at least some of the specific details of the link between bodily and moral purity are culture-specific. For instance, facial purification has especially strong moral effects in East Asian cultures that connect the face and public image to a higher degree than other cultures. In individuals with East Asian cultural backgrounds, face-cleaning (but not hand-washing) decreases feelings of regret and guilt and liberates them from prosocial behavior motivated by guilt (Lee et al. 2015).

Finally, complex scaffolding modifies a relatively strong metaphysical claim in Lakoff and Johnson's work. The claim is that during the previously described period of "conflation," stable neural connections are generated that shape conceptual domains, leading to the conclusion that "neural mapping circuits ... constitute the metaphor" (Lakoff 2008, 26; 2014). But the notion of constitution is ambiguous and appears to indicate an excessively strong metaphysical relationship. Instead, it is better to understand this claim in a more relaxed manner—namely, as the metaphysical claim that metaphors are *realized* in neural circuits that link a source region to a target region in a way that gives rise to uni- or bidimensional cross-domain effects. Further, the studies offer no indications that conceptual metaphors are exclusively or necessarily realized in brain circuits. For this reason, complex scaffolding aims to remain compatible with extended and enactivist accounts, which would perhaps claim that metaphors could at least partially be realized in parts of the body or the environment (see Varga 2018b).

Overall, complex scaffolding—within the ASC framework and supported by the behavioral and neuroscientific studies—offers a promising account that may have interesting implications for studying aspects of mental disorders. Although unanswered questions remain, we may suspect that the effects between sensory motor and higher-order domains might be exploited for healthcare-related applications and translated into clinical practice.

6.4 Complex Scaffolding, Depression, and Psychomotor Retardation

Cross-cultural psychiatric research shows that individuals in different cultural contexts tend to report different bodily symptoms, but it is widely acknowledged that in some form, nearly the entire range of mental disorders includes altered bodily experience. For instance, individuals with depression describe that their bodies, which under ordinary circumstances operate as the silent medium through which the world is experienced, are elevated to the center of experience and become uncomfortably conspicuous and object-like (Fuchs 2002, 2005). Affected individuals frequently report feeling passively at the mercy of physical forces like gravity and deploy metaphorical constructs like "being down," "being pressed down," "weighed down," and "descent" (McMullen and Conway 2002; see also Kövecses 2005). Such constructs enable patients to comprehend and coherently narrate the experience of what psychiatrists and psychologists refer to as "psychomotor retardation." The term refers to alterations in a wide range of actions, including speech, motility, fine motor behavior, gross locomotor activity, and facial expression. It includes reduced speed (of legs, hand, torso, and head movements), reduced arm-swinging and upper-body movements, slumped posture, and specific alterations in gait patterns (for comprehensive reviews, see Sobin, Mayer, and Endicott 1998; Michalak et al. 2009; Buyukdura, McClintock, and Croarkin 2011; Bennabi et al. 2013). Changes in gait pattern are in these cases not attributable to reduced walking speed and include smaller amplitude in arm-swinging and upper-body vertical movements, slumped and forward-leaning posture, and a larger amplitude in lateral body swaying (Michalak et al. 2009).

Besides being included among the DSM-5 (2013) symptom criteria required for a diagnosis, psychomotor retardation is correlated with depression severity (Lemke et al. 2000) and has proven to be of both diagnostic and prognostic value (Calugi et al. 2011). The intensity of psychomotor retardation predicts delayed response to treatment with psychotherapy (Frank et al. 2011) and is correlated with a decrease in cognitive performance (Lemelin and Baruch 1998; Caligiuri and Ellwanger 2000) in terms of attention, visual learning, memory, and executive function (i.e., verbal fluency, attentional and cognitive flexibility). For instance, in visuospatial cognitive operations, measured by mental ego-rotation and object-rotation tasks, patients exhibited specific deficits compared to controls (Chen et al. 2013). In severely depressed patients undergoing repetitive transcranial magnetic stimulation treatment, a significant cognitive improvement covaried with the improvement of psychomotor retardation (Ullrich et al. 2012), whereas positive effects of pharmacological treatments, measured by depression-rating scale scores, correlated with changes in motor performances (Sabbe et al. 1996).

6.4.1 Psychomotor Retardation as a Symptom

Psychomotor retardation thus is a crucial feature of depression, and it is now acknowledged that it is also useful in distinguishing some depressive subtypes (Parker 2007). Nonetheless, the nature of the exact relationship between depression and psychomotor retardation is surprisingly under-researched. The literature mostly regards psychomotor retardation as a *symptom* or sign of depression (see, e.g., Schrijvers, Hulstijn, and Sabbe 2008). Influential cognitivist psychotherapeutic approaches tend to go further, often claiming that psychomotor retardation is in some way caused by the depressive cognitive style and the ensuing emotional changes. Based on a cognitivist framework, Beck (1995, 37–39) describes how the depressively biased core belief "I'm incompetent" reverberates through the cognitive system, causing automatic thoughts about worthlessness and leading to feelings of discouragement, together with "heaviness in the body." Thus, somewhat crudely put, much like psychosis is a symptom of a psychotic disorder or skin rash is a symptom or sign of a viral infection, psychomotor retardation is a symptom or sign of poor motivation and low mood that ultimately results from cognitive biases.

Using the terminology introduced in chapter 2, we can say that the clinical cognitivist account of depression subscribes to both ontological and epistemological separability theses, or at least something in their close vicinity. This has a couple of implications. First, if psychomotor retardation is merely the symptom or sign of the disorder in the earlier mentioned sense, then it does not constitute a proper part of it. Second, although psychomotor retardation might be useful for diagnosis and prediction, once we have comprehended the cognitive and motivational aspects of depression, a detailed account of psychomotor retardation is not mandatory. Third, if psychomotor retardation is a symptom or sign of depression, indicating a unidirectional causal relation, then trying to reverse it—for instance, by nonbiochemical interventions—will not likely be effective to alleviate depression. It would be like treating the rash in order to treat the viral infection.

6.4.2 Depression, Psychomotor Retardation, and ASC

The clinical cognitivist framework, however, runs into significant challenges and fails to account for recent findings. Studies indicate that bottom-up interventions decreasing specific features of psychomotor retardation, such as changing posture and gait condition, can alleviate symptoms and lead to cognitive and emotional changes. For instance, a recent study manipulated gait patterns in a nonclinical sample. Participants, who had no knowledge about the intention of the experiment, were given computer-based feedback to alter their gait patterns to adopt a depressed or happy walking style. The authors investigated the processing of self-referent material by measuring the numbers of positive and negative words that participants recollected from a list presented to them earlier (Michalak, Rohde, and Troje 2015). Compared to participants in the happy condition, participants in the depressed condition recollected more negative self-relevant material, leading the authors to conclude that "biased memory towards self-referent negative material can be changed by manipulating the style of walking" (Michalak, Rohde, and Troje 2015, 124).

One might be inclined to think that the effect is indirect, in the sense that the changed gait pattern induces an emotional change, which subsequently affects the processing of self-relevant material. However, the authors found no indication for affective changes and conclude that "the induced changes in affective memory bias were directly caused by somatic processes" (Michalak, Rohde, and Troje 2015, 124). Although these findings indicate that relatively brief gait changes affect subtle cognitive processes related to memory, another recent study demonstrates that posture impinges on memory bias in inpatients with depression. The participants, psychiatric inpatients with depression who scored more than 14 on the Beck Depression Inventory-II and were unaware of the purpose of the experiment, were asked to sit either in a slumped posture (depressed condition) or in an upright posture (nondepressed condition). To investigate the effects of the postures on the way in which self-referent material is processed, positive and depression-related words were presented one by one to the participants. After completing a subsequent distractor task, they were asked to recall these words. In comparison with upright-sitting participants, participants in the slumped condition recollected fewer positive than negative words. Also, participants in the upright condition exhibited a balanced recall, recalling fewer negative words and more positive words (Michalak, Michnat, and Teismann 2014).

Although more research is needed to corroborate these findings, especially in clinical settings, they are consistent with other results showing that different postures and facial expressions modulate the latencies in retrieving positive vs. negative autobiographical memories. In contrast to the speed of recollecting negatively valenced memories, having an erect posture and smiling (as opposed to slumped posture and frowning) increases the speed of recollecting pleasurable memories (Riskind 1983). Correspondingly, the recall of positive events is slower and more difficult to achieve in a slumped posture than in an upright posture, whereas adopting an upright posture increases access to positive content, reduces fatigue in depression (C. Wilkes et al. 2017), and helps maintain self-esteem, increased positive mood, and speech rate during the Trier Social Stress Test (Nair et al. 2015).

The arising conundrum for proponents of the separability thesis is now quite palpable. If we adopt the view that psychomotor retardation and higher-level cognition are at least epistemologically inseparable, then the results of these studies are quite straightforward and predictable. If, however, we adopt the separability thesis and consequently regard psychomotor retardation as a symptom of depression, then these findings look mysterious and difficult to explain. Put crudely, an explanation within the latter framework would require accepting something like the dubious view that sometimes treating the rash can treat the viral infection.

It is important to add that the results cannot be explained in terms of simple scaffolding. We may think that there is some kind of affinity between changing the activity of the corrugator muscle due to Botox injection and changing gait pattern and posture. Both affect the body, and given that cognitive activity is scaffolded onto the sensorimotor domain, it makes sense to think that gait patterns and posture will exert bottom-up influence on cognitive activity. However, simple scaffolding does not offer a comprehensive story. In contrast, complex scaffolding clarifies the relationship of the conceptually linked bodily and cognitive aspects to an extent that could guide further reflection on possible therapeutic interventions.

6.4.3 Clinical Implications

Before offering some reflections on possible practical implications, it is fitting to emphasize that the evidence needs to be strengthened before substantial conclusions can be reached. Future studies would benefit from operating with larger sample sizes and from further investigating details of the relevant effects by including nondepressed control groups, as well as a condition in which the sensorimotor system is not manipulated. Also, much of the available literature is limited to bottom-up influences, and the overall framework could be strengthened if future studies could provide evidence for top-down effects in depressed patients.

Possible clinical applications remain to be explored and empirically tested, but the fact that complex scaffolding can explain the findings suggests that it might provide insight to improve current approaches. More precisely, the claim is that it may improve the overall understanding between mental health professional and patient, extend the range of diagnostic options, and point toward additional possibilities for therapeutic interventions. Let us briefly review these possibilities.

One positive potential consequence is an *enhanced understanding* of the patient's overall condition. An approach that highlights the body will likely specifically benefit many patients who are comfortable with talking about bodily disturbances, but less so with conveying emotional disturbances and disquieting thoughts. It would give the mental health professional a framework to organize seemingly separate bodily, emotional, and cognitive states into a comprehensive picture in which the cognitive biases, emotional "slumps," and slumped posture reflect different sides of the same experience. The focus on the nature of bodily experience could provide the professionals and patients with a window into the individual "profile" of the condition and contribute to an enhanced understanding, communication, and alliance. In comparison, therapists subscribing to a broadly cognitivist approach will regard complaints about motor retardation as distinct from emotional and cognitive changes and as epistemologically less valuable for understanding a condition. They therefore might try to shift the focus from the patient's bodily experiences to details of her disturbing emotions and cognitions, which may mean losing an important source of insight. This would be unfortunate, as altered bodily experience does not merely reliably accompany depression but is sometimes its most salient feature.

Utilizing complex scaffolding as a theoretical framework could form the basis for developing a *fine-grained taxonomy of bodily changes*, which could then be used for diagnostic purposes. A more precise understanding of the range and nature of bodily changes could help *identify psychomotor markers* that are conveniently obtained in a clinical setting and amenable to objective assessment through simple motor tasks. This requires conceptual work, careful attention to the phenomenology of bodily experience, and establishing how particular bodily aspects are related to cognitive and emotional phenomena. But it could lead to promising advances. In fact, the diagnostic role of psychomotor symptoms is showing initial promise, and it appears useful for differential diagnosis and subtyping depressed patients into melancholic and nonmelancholic groups (Pier, Hulstijn, and Sabbe 2004).

The most often used clinical rating scales for depression will, however, not provide much assistance going forward. They typically include only one item for psychomotor disturbance and are not fine-grained enough to differentiate retardation from fatigue, loss of energy, or lack of concentration. The perhaps most detailed scale designed to assess psychomotor retardation in depression is the Salpetriere Retardation Rating Scale (SRRS). The SRRS is particularly interesting in this context. It has had success with predicting responses to antidepressant medication, and it focuses on cognitive and motor aspects of retardation, assessing gait, limb movement, speech, and cognitive function. However, it is also the case that the SRRS would benefit from integrating more objective measures that are currently scattered across the entire range of research. For instance, to increase fit with the particular profile of depression in a patient, the SRRS could be combined with measures of alterations in paralinguistic aspects of speech (fluency, prosody, and speech pause time; Flint et al. 1993; Cannizzaro et al. 2004) and kinematic analysis of drawing and handwriting movements (Pier, Hulstijn, and Sabbe 2004; Mergl et al. 2004).

Bringing to light the body and the motor system as promising *loci of therapeutic interventions* would expand the range of therapies offered to patients. Complex scaffolding could provide the background to design undemanding and low-risk interventions on the body, which could be applied either by themselves or alongside other treatments. Such body-based interventions would be ideal for diverse populations of patients and would particularly benefit those who are not proficient speakers or who, due to personality or cultural barriers, have a difficult time conveying their thoughts and feelings. Of course, the long-term success of such interventions will to a decisive degree depend on the extent to which changes in motoric patterns have a lasting impact. This question can be explored in longitudinal studies. Optimally, if patients with depression are successfully trained to enduringly alter dysfunctional motoric patterns, then the habits of the body may become habits of the mind, providing a lasting shield against cognitive and emotional vulnerability to processing biases.

6.5 Conclusion

Based on evidence that cognition is scaffolded onto the sensorimotor system, this chapter explored complex scaffolding, a scaffolding relationship characterized in terms of *Active Scaffolding (MM)*. It provided support for the overall ASC framework, explored several uni- and bidirectional effects, drawing attention to possible contributions for a more comprehensive understanding of mental mechanisms and potential therapeutic implications. Clearly, systematic clinical application to depression would require both a comprehensive grasp of the details of psychomotor retardation and a more precise chart of uni- and bidirectional effects. ASC offers the theoretical background for such a project, and although significant challenges remain, embarking on such a project seems attractive in light of the numerous potential benefits for diagnosis and treatment.

But this should not hide from view that the nature of the empirical material and the gap between the theoretical framework and the kind of interventions the studies deploy are such that compromises are difficult to avoid. It is not just that active scaffolding is sometimes inferred from complementary but separate studies. It is sometimes even difficult to determine how the studies should be evaluated because it is not clear whether they intervene in a top-down or a bottom-up manner. For example, in the case of physical cleansing, is it the perception of cleanliness or the relevant belief that is the influencing factor? Such considerations should be kept in mind when assessing the conclusions of this chapter.

7 Intersomatic Scaffolding: Being in Sync

7.1 Interaction and Intersomatic Scaffolding

This book has thus far mainly been concerned with forms of intrasomatic scaffolding. Expanding the scope of inquiry, this chapter addresses a particular form of simple scaffolding in which the scaffolding not only includes the agent's sensorimotor apparatus, but extends into the environment to include external structures that adaptively guide behavior. In such cases, the vehicles of cognition individuate externally and are best described as assemblies that include parts of the environment along with the brain and the body. This chapter is thus dedicated to investigating a form of *intersomatic scaffolding*, in which the vehicles of cognition include the body of another human being. As in previous chapters, the goal is to offer reflections on how a deeper grasp of these scaffolded processes could contribute to a more complete comprehension of characteristic disturbances.

To obtain a proper grasp on the subtleties of these cognitive processes, it is helpful to begin by exploring the codevelopment of interactive skills in infants and, in particular, synchronic interaction patterns between infant and caretaker. As we will discuss, these are not abilities that are transcended and replaced by more complex ones in the course of development; they remain operational and scaffold cognitive processes in adulthood. The bulk of this chapter will address emotion regulation, a task that can be accomplished by a particular type of embodied interaction that has its roots in infancy. Emotion regulation is particularly interesting for our goals because it is an ability that can be scaffolded in various social and nonsocial ways, and it is often impaired in mental disorders. Although we will illustrate how emotion regulation can be scaffolded onto the body, the focus will be on how the task of regulation can be scaffolded onto face-to-face interactions, such that the machinery that realizes this process includes parts of the body of another human being, as well as the bodily interaction itself. In such cases, emotion regulation is scaffolded onto *synchronically* interacting bodies.

7.2 Emotion Regulation

Emotions, roughly defined as matched arrangements of responses to internal or external changes, are not only crucial features of human experience but have a particular significance for the organism from an evolutionary perspective. They can be seen as fulfilling fast processing tasks, evolved to enable rapid adaptive responses without requiring considerable cognitive effort (Tooby and Cosmides 2008). Moreover, they serve important social and cultural functions, prepare the body for action, calibrate cognition, and coordinate competing systems to offer adaptive reactions (Damasio 1994).

Of course, emotions and the response tendencies they give rise to can be inappropriate, and unrestrained emotional impulses often interfere with our ability to engage in optimal decision-making and adaptive behavior. Exerting some level of regulatory control over emotions is therefore indispensable to flexibly adapt to situational demands. "Emotion regulation" denotes processes that are responsible for "monitoring, evaluating, and modifying emotional reactions" (Thompson 1994, 28).[1] This often involves decreasing and increasing the intensity magnitude, duration, onset, and offset of emotions. Human beings typically attempt to increase positive emotions that they want to preserve or fail to display appropriately (not happy enough to see a parent) and decrease emotions when they prompt maladaptive behavioral responses (shouting at an annoying student) or cause psychological suffering.

Emotion regulation is particularly interesting for our goals. The ability to effectively regulate undesired emotions and to modify the course of affective responses is crucial to maintaining mental health (Gross, Richards, and John 2006; Berking and Whitley 2014), and this ability is impaired in most of the mental disorders included in the DSM-5 (2013). For example, individuals with depression not only suffer from biases that cause negative emotions, but also fail to effectively regulate and modify these emotions (Ehring et al. 2008; Kassel, Bornovalova, and Mehta 2007). These individuals display a decreased use of adaptive strategies such as cognitive reappraisal (see Aldao, Nolen-Hoeksema, and Schweizer 2010) and use maladaptive strategies like ruminative elaboration and expressive suppression, which influences distress and other symptoms (Nolen-Hoeksema, Wisco, and Lyubomirsky 2008; Aldao and Nolen-Hoeksema 2012a, 2012b). Effective emotion regulation is negatively correlated to depression and can generally assist in maintaining mental health.

Adult human agents have several strategies at their disposal to regulate emotion and to manage emotionally arousing experiences (see Gross 2014; Gross and Thompson 2007; Gross, Sheppes, and Urry 2011). It can be affected by conscious, effortful, and controlled processes, but also by several unconscious, effortless, and automatic processes. One obvious way to avoid some unwanted emotion is to avoid situations and environments that typically give rise to the emotion or that are highly likely to trigger the emotion. It is, however, debatable to what extent this constitutes emotion *regulation* because what one regulates is primarily the likelihood of its formation. But when it comes to a more direct regulation of occurring emotions, there are at least three main strategies that we commonly make use of:

(1) *Cognitive change* involves reappraising the situation to alter its emotional impact, by reinterpreting either the situation or our capacity to manage its demands.

(2) *Response modulation* includes motor-behavior modifications, influencing physiological or behavioral response to the emotion—for instance, by downregulating expressive behavior, which tends to increase the salience of the emotion.[2]

(3) *Attentional deployment* involves decreasing stimulation by withdrawing and redirecting attention away from the emotion-eliciting stimulus.

Besides the evident link to mental disorders, what makes these emotion-regulation strategies interesting for our goals is that they illustrate how a cognitive task can be offloaded and achieved internally and externally in several ways. Ahead, the aim is to use modified versions of Inga-and-Otto types of situations to illustrate how this scaffolding allows extension into the environment.

7.2.1 Extended Emotion Regulation

Over the course of the last year, Inga and Otto both experience major setbacks in their lives, and they now meet at least two criteria for depression, an almost constant state with depressed mood and "feelings of worthlessness." To their therapists, they describe these as one all-encompassing emotional state, the management of which consumes all their energy. To regulate the unwanted affective state, they can deploy three strategies:

(1) *Cognitive change:* Inga and Otto have learned in their therapy sessions that an effective way of dealing with a painful and undesirable emotional state is to reappraise the situation and the origin of the state. Inga, in full command of her cognitive abilities, consults her memory and recalls a list she had formulated with the therapist that unambiguously affirms her worth by highlighting all her qualities and accomplishments that the depression tends to overshadow. Otto, who suffers from memory-related cognitive deficits, consults his notebook, which contains a list he had formulated with his therapist. The list highlights all his qualities and accomplishments in exactly the same way as Inga's "mental" list.

(2) *Response modulation:* Inga and Otto have received advanced psychoeducation and know that changing physiological and behavioral expressive patterns may be used to regulate emotions. They understand that regulation can be achieved by altering various peripheral physiological responses (heart rate, muscle activity, breathing) or by changing expressive behavior and proprioceptive input, through changing posture, gait, facial muscle activation, head movement, and isometric muscle contraction in the arms (for a review, see Shafir 2015). Knowing how to exploit what we have referred to as *simple scaffolding structures*, they attempt to regulate their states by diminishing facial muscle activation. Inga, who is in full command of her body, closes her eyes, focuses on the relevant facial muscles (like the corrugator muscle in the glabellar region), and effectively relaxes them. Otto, who is not in full command of his facial muscles and unable to alter their activity as Inga can, uses a Botox injection to temporarily relax the relevant muscles.

(3) *Attentional deployment:* During their sessions, Inga and Otto learn that our selective attention mechanism that filters sensory stimuli can be

employed in the service of emotion regulation. Once trained to selectively attend to particular stimuli, this regulatory strategy only requires minimal cognitive effort and works by reorienting overly self-directed and negatively biased attentional patterns toward more neutral content. Inga successfully trains to correct attentional biases and quickly disengages from negative stimuli to refocus attention on neutral stimuli. She thereby alleviates her low mood and diminishes the impact of negative emotions. Otto, who is unable to achieve this due to his physical and cognitive impairments, has a device attached to his body that monitors negatively biased attentional patterns and produces a signal that leads to the reorientation of Otto's attention toward more neutral cues.

These thought experiments are somewhat rudimentary, but they nevertheless indicate the range of internal and external possibilities for emotion regulation. Importantly, in all these cases, Inga's and Otto's processes are cognitively equivalent, playing identical functional roles. If we look closer at the second strategy, we see that Inga's response modulation involves manipulating her body, exploiting the intrasomatic scaffolding of emotions. In the remainder of this chapter, the focus will be on how the task of regulation can be achieved through intersomatic scaffolding. It will be shown that emotion regulation can be accomplished in dyadic bodily interactions, such that the vehicles that realize this process include parts of the body of another human being. Such emotion regulation is scaffolded onto not a single body, but two synchronically interacting bodies.

7.3 Synchrony

Synchronization is a widespread phenomenon, occurring on multiple systemic levels in nature, technology, and social interaction. Whether in physics, chemistry, engineering, medicine, or social sciences, synchronization is understood as the coupling of entities to form a common behavioral organization, and it is seen as a manifestation of a characteristic tendency toward self-organization in nature (Pikovsky, Rosenblum, and Kurths 2001; Osipov, Kurths, and Zhou 2007). Synchronization phenomena have been investigated at least since the seventeenth-century discovery that pendulum clocks attached to the same support spontaneously synchronize.

Since then, researchers have explored a wide variety of dynamic synchronizations in flocks of birds, schools of fish, and chirping of crickets, but also in human interactions. In all these instances, synchrony may be guided by analogous self-organizing principles (Kelso 2009; Schmidt and Richardson 2008).

Interpersonal behavioral synchrony between human beings belongs to natural synchrony phenomena, whereas the characteristic coupling it exhibits is consistent with self-organized coordination dynamics (Miles et al. 2010). In this context, synchrony refers to behavioral coordination in interactions, which is nonrandom, patterned, and matched in both timing and form (Bernieri and Rosenthal 1991). Importantly, such synchrony is reducible to neither behavioral mimicry[3] nor to joint action. The latter involves individuals who aim to accomplish a task by time-coordinated joint activity, which need not be imitative. The former, in which the observed behavior of the interaction partner involuntarily increases the probability of displaying the same behavior, is an unconscious and automatic one-way process. Although it produces behaviors that are similar in form and close in timing, it does not require anticipating the other person's behaviors. In contrast, the synchronization of matching behaviors shares features of both coordination and imitation, displaying sensitivity to the behavior of the other without there being a concrete affiliative or coordinating task (Louwerse et al. 2012). As Feldman (2017, 87) puts it, synchrony unfolds as "a stochastic process, follows dynamic systems' principles, and integrates patterned order and local variability."

Synchronous interactions arise during early periods of infant-caretaker interactions but continue throughout life and can be observed in everyday communicative interactions between adults—in particular, those in close relationships (see Wheatley et al. 2012; Delaherche et al. 2012). When interacting, for instance, in everyday face-to-face conversations, without apparent cognitive mediation or conscious awareness, a synchronic alignment of body postures occurs (Shockley, Santana, and Fowler 2003; Varlet et al. 2011). People spontaneously synchronize movements when they walk together (van Ulzen et al. 2008), clap (Néda et al. 2000), form waves (Lakens 2010), rock on chairs (Richardson et al. 2008), swing pendulums (Schmidt and O'Brien 1997), or tap fingers (Oullier et al. 2008). Also, without deliberate intent, speakers have a strong tendency to adopt the syntactic form used by their conversational partner (Branigan, Pickering and

Cleland 2000; Dale and Spivey 2006), as well as to coordinate speaking rate, acoustic and prosodic features (Levitan and Hirschberg 2011), and eye movements (Richardson and Dale 2005). Although many behavioral features may become synchronized, many of these synchronizations cannot be directly perceived. However, studies using electromyography have shown that participants display muscle-activation patterns that are unobservable to the naked eye and that match the facial expressions of emotions that the participants were observing (for reviews, see Hatfield, Rapson, and Le 2009; Lundqvist 1995).

7.3.1 The Regulatory Function of Synchrony

Behavioral synchrony can take on different roles, but its relatively widespread occurrence offers preliminary reasons to assume that it has some kind of function linked to perception-action coupling across interlocutors. This is not an exhaustive list of possibilities, but there are two regulatory functions that are worth elaborating on.

The first is perhaps best seen as a *group-level function*, in which the synchronization of behaviors with interaction partners is a functional part of communication (Louwerse et al. 2012) and increases connectedness, liking, and social bonding (Chartrand and Lakin 2013; Wheatley et al. 2012; Feldman 2017), which most likely augments social coherence. For instance, participants tapping in synchrony with the experimenter later judged the experimenter as more "likeable," whereas interactions without synchrony received less positive scores (Hove and Risen 2009; Marsh, Richardson, and Schmidt 2009). Moreover, synchrony promotes prosocial attitudes, cooperation, and helping behavior (Bernieri et al. 1994; Mazzurega et al. 2011; Wiltermuth 2012; Kirschner and Tomasello 2010) and provides an efficient mechanism for assisting social understanding (Marsh, Richardson, and Schmidt 2009; Semin and Cacioppo 2008; Schmidt and Richardson 2008), which may explain why individuals may compensate for their observed differences by attempting to increase bodily synchrony (Miles et al. 2010).

To illustrate the enhanced sense of proximity to other human beings that it provides, it is helpful to invoke a study on participants who are brushed on the cheek while watching a video of a stranger being brushed in the same manner, either in synchrony or asynchrony (Paladino et al. 2010). In the synchronous stimulation condition, but not in the asynchronous condition, participants believed that the face and personality of the viewed person resembled them more, and they tended to confuse whether they have felt or seen the sensations.

On the individual level, the social proximity to conspecifics and synchronization of behaviors with interaction partners has a *self-regulatory function*. In general, synchronization reduces the cognitive load by "constantly computing the next behavior in each of classes 1 to some large n during a task" (Louwerse et al. 2012, 1422). Such cognitive offloading, with multiple channels synchronized within the same interaction, may support cognition by decreasing comprehensive planning requirements for each channel (gesture, posture, language, etc.) during interaction (Louwerse et al. 2012; Garrod and Pickering, 2009). Consistent with the tendency to offload cognitive tasks, we enter into psychological interdependencies with trusted fellow human beings and use their resources to restore our emotional balance via overt behavior (Coan 2010; Coan and Maresh 2014). Compared to cognitively achieved self-regulation, which is costly and drains cognitive resources like working memory, solving psychological regulatory tasks such as emotion regulation through simple social proximity comes at a very limited cognitive cost. Although it is not entirely clear whether this is a consequence of regulative capacities, interpersonal synchrony affects self-evaluations: being prevented from synchronizing with others tends to decrease self-esteem, whereas a lot of intentional synchronous movement with a confederate boosts self-esteem (Lumsden, Miles, and Macrae 2014).

This is not to suggest that there is a single stable mode in which synchrony unfolds. Rather, on different levels of interaction (pitch, motor behavior, linguistic coordination, etc.), there might be different patterns of synchrony, with precise pattern details depending on task-based modulation. Furthermore, it is worth highlighting that despite numerous studies showing synchronization of behavior across different channels, much remains undetermined about the precise temporal structure in synchronizations or about how channels may be synchronized differently during the same face-to-face interaction.

7.3.2 Interactional Synchrony and Emotion Regulation

Synchronization thus in many cases can be seen as an adaptive problem-solving capacity, and it is relevant to multiple functions. To provide a fuller understanding of how it works, it is useful to explore some interactive capabilities that emerge in infancy. Although numerous terms have been used in the literature to describe patterns of interdependence in infant-caretaker behaviors (social resonance, coordination, attunement, etc.), *synchrony* effectively highlights the smooth exchange of temporally matched movements, vocalizations, imitation, and mirroring between the child and the caretaker. But such synchrony in the developmental context also has diachronic functions, which is why we may define it as "the temporal coordination of micro-level relational behaviors into patterned configurations that become internalized and serve to shape infant development over time and repeated experience" (Leclère et al. 2014, 4; Harrist and Waugh 2002).

An early form of synchrony is noticeable in newborns mimicking the facial behavior of their caregivers. This type of coordination requires perceiving, understanding, and flexibly adapting to communicative signals, and newborns are already equipped with some basic capacities that make it possible. Neonatal imitation and other paradigms in developmental psychology research point to the existence of an innate intermodal connection between perception and proprioceptive awareness, which allows infants to connect and synchronize with others. For instance, newborns as young as one hour old can imitate facial expressions that they obviously have not seen themselves do (Meltzoff and Moore 1983, 1989). Throughout their early development, their ability to comprehend cross-modal correspondences between perception and proprioception enables them to synchronize tongue protrusion, hand, head, lip, and cheek movements, eye blinking, and mouth opening (Heimann 2002; Meltzoff 2005; Meltzoff and Moore 1997).[4]

Such synchronizations are often intermodal and in many cases not obvious to the naked eye. However, refined microanalytic technologies and measurements have aided uncovering a variety of synchronic organization in face-to-face interactions. Studies measuring facial muscle activity in five-month-old infants demonstrated rapid-matching facial mimicry in response to perceiving others' emotional expressions (Isomura and Nakano 2016).[5] Other technologies have helped reveal matching micro-level body movements, gaze, vocalizations, and affective behaviors (Trevarthen 2002; Weinberg and Tronick 1994; Feldman and Eidemann 2004). Such technologies have facilitated understanding that synchronic interaction involves not only mirroring, but also the support and completion of the gestures, movements, and vocalizations of the other (Feldman 2007, 2017). This allows progressive changes and evolving patterns of "narratives" in which the dyad progresses together toward higher or lower levels of arousal. These narratives can unfold in different ways, with mother-infant synchrony and father-infant synchrony exhibiting different temporal and dynamic structures (Feldman 2003). Also, it has become clearer that synchrony need not involve perfectly symmetric timing. Interruptions and deviations are in fact important to improving adaptation and creativity (Leclère et al. 2014), and the regular "reparation" of mismatched states appears to be more important than absolute matching of positive affect (Beeghly and Tronick 2011; Tronick and Beeghly 2011; Ostlund et al. 2017). If some of these interactions are mismatched, infants become distressed and seek resynchronization (Tronick, Als, and Adamson 1979; Tronick 1989; Murray and Trevarthen 1986; Reck et al. 2004).

Neonates and infants have precursory strategies to regulate emotions. For instance, to control the level of arousal, they may engage in self-soothing behavior (thumb-sucking) or withdraw from an emotion-eliciting stimulus. It is more accurate, however, to comprehend these strategies as more or less automatic reflexes. This is different from cases in which the emotional arousal of the infant is accomplished by the mutually coordinated synchronic interaction between infant and caretaker. Furthermore, synchronic interaction not only drives real-time emotion regulation, but also progressively builds self-regulation capacities and enhances children's capacity for emotion regulation in the physical absence of the caretaker (Tronick 2002; Feldman 2007, 2015). Correspondingly, the lack of sustained exposure to synchrony has both short- and long-term effects on the infant's emotion regulation. For instance, interactions between infants and mothers suffering from depression are poorer and exhibit lower levels of synchrony and less sharing of positive affective behaviors (Weinberg et al. 2006; Moore and Calkins 2004; Zlochower and Cohn 1996). This may be considered analogous to a bottom-up intervention. There is evidence that these infants have difficulties with effectively regulating emotions and acquire maladaptive regulation that can lead to long-lasting averse psychological effects (Reck et al. 2004; Manian and Bornstein 2009; Tronick and Gianino 1986).

Synchronic nonverbal behavior and its regulatory effect on emotions is not restricted to early development. Although adult dyads exhibit interpersonal synchrony, as well as emotional contagion and convergence (Butner,

Diamond, and Hicks 2007), the literature on emotion regulation in adults primarily has focused on intrapersonal processes or social influences in close relationships that aim to support self-regulation (providing positive stimuli, interrupting rumination, promoting physical activity, giving advice, soothing, etc.). Less attention has been payed to forms of synchrony with regulative function in adult relationships, but it seems unlikely that an effective form of dyadic emotion regulation throughout development would completely cease to exist in adulthood (Rimé 2009). Instead, there is what chapter 3 has referred to as *ontogenetic integration*. Consistent with the tendency to offload and reuse, attachments later in life to close friends, romantic partners, or various in-group associates "repurpose the basic machinery established by the mother-offspring bond during early 'sensitive periods'" (Feldman 2017, 81). For example, the mutual regulation of each other's psychophysiological arousal is a significant aspect of romantic attachment in adults (Sbarra and Hazan 2008). Couples' emotional and related physiological states influence each other without conscious awareness (Saxbe and Repetti 2010), and one's close partner may sometimes act as an "external drive" for emotion regulation and counterbalance deficits associated with psychopathology (Marroquín and Nolen-Hoeksema 2015). Similar to the infant-caretaker dyad, emotion regulation in adult couples does not necessarily require mirroring, but rather the balancing of affect and arousal. This is consistent with the fact that high degrees of coordination do not necessarily indicate adaptive emotion regulation but may instead indicate relationship distress.

7.3.3 Intersomatically Scaffolded Emotion Regulation

Before we get to the specific implications of scaffolded emotion regulation for mental disorders, it is important to identify the appropriate theoretical framework for this phenomenon. Developmental psychologists have in fact used the term scaffolding to describe synchronically organized processes and to maintain that the caregiver offers the infant a scaffolding for emotion regulation (see, e.g., Bruner 1975; Stern 1985). Developing this thought, instead of assuming a unilateral, sequential adjustment of one partner to the other, some argued that the caretaker and infant dyad is best seen as constituting a system that regulates the infant's emotional states (Tronick 1998; Feldman 2007). On such a basis, the phenomenon appears to be adequately describable in terms of extended cognition. Translated into the vocabulary used in debates on extended cognition, the vehicles that realize the emotion regulation of the infant extend beyond the limitations of her body and incorporate external resources made available by the caretaker. Much like the integrated system that Otto forms together with his notebook to solve the memory task, the infant and caretaker form an integrated system to solve the emotion-regulation task. This system is functionally equivalent to the normally operating emotion regulation of a single individual, achieved, for example, by cognitive reappraisal or emotion inhibition.

It is possible to raise several objections to this conclusion. One might insist that the synchronic interaction merely offers a causal albeit crucial contribution to emotion regulation. However, this is contestable. We may very well be dealing with entities in a state of "continuous reciprocal causation" (Clark 1997; Wheeler 2005) such that the behavior of the caretaker and infant simultaneously affect the behavior of each other in a dynamical manner. But then, explaining the interaction in terms of separate causally active inputs and outputs oversimplifies the complex mutually modulatory dynamics (see Varga 2016).[6]

But we may also ask whether the extended cognition label is justified here. The infant-caretaker interaction and the Otto-notebook interaction appear quite different in several respects. There are at least four such differences worth highlighting:

(1) Unlike in the case of Otto, it would be incorrect to claim that the infant is engaged in some kind of epistemic action or that the infant manipulates the relevant resources of the caretaker to regulate her emotions. Moreover, it would be odd to say that the "information" received from the caretaker is automatically endorsed. Instead, the infant merely engages in a pleasurable interaction that leads to the formation of a cognitive system.

(2) In contrast to Otto's notebook, the caretaker is not constantly available, and the interaction is not reliable. Indeed, as we have seen, it is unstable and prone to desynchronization.

(3) Otto is in control of the retrieval process and has at some past point consciously chosen to use an external entity to accomplish a fixed task. In contrast, the dyadic process is not controlled by either part of the

system, and emotion regulation *emerges* as the unintended by-product of interaction.

(4) Whereas the belief about the location of the museum that arises in the Otto-and-the-notebook coupled system is not novel compared to what is entailed in the parts of the system (it is in the notebook), emotion regulation is irreducible and not the simple linear "sum" of the infant's and the caretaker's states. At least at a certain developmental stage, it is a diachronically novel *irreducibly emergent* systemic property, instantiated in the system rather than a single human agent, without following from the behavior of its parts (for a more detailed discussion, see Varga 2016).

That said, it is critical to observe that not all cases of socially or even dyadically scaffolded cognition exhibit such characteristics. To see why, imagine that instead of his notebook, Otto could be using the intact memory of his wife, with whom he had shared most of his adult life. Otto consulting his long-term partner about the location of the museum could count as extended cognition but would lack the dynamic characteristics and exhibit an aggregative decomposition. In this sense, the dyadic emotion regulation is more reminiscent of certain groups that display irreducible cognitive capacities (see, e.g., Theiner, Allen, and Goldstone 2010), but without the common goal and knowledge about the domains of expertise of other individuals that is involved in group cognition.

7.4 Two Clinical Implications

This account of intersomatically scaffolded emotion regulation with emergent, system-level cognitive properties can be deployed to shed light on processes in psychotherapy. Emotion regulation is a critical goal in psychotherapy, and though conventional cognitive approaches have particular methods to assist in achieving this goal, they are hard to apply when patients are overwhelmed by intense and fluctuating emotion, characterized by disproportionate intensification (dread, rage, panic, sense of urgency, etc.) or excessive deactivation (dissociation, depersonalization, emotional numbing, etc.). Utilizing cognitive techniques or behavioral exposure may even stimulate further stress (Leahy, Tursch, and Napolitano 2011, ix). For this reason, some level of intact emotion regulation is necessary for being able to benefit from cognitive interventions.

Recognizing this issue, approaches like *emotional schema therapy* first focus on providing assistance to express, label, validate, accept, and contextualize emotional experience. With the aim to avoid suppression or elimination, the goal is to normalize and regulate emotions by finding constructive ways to use them—for instance, by conveying that such experiences may offer a chance to form closer relationships (Leahy 2009). However, the focus in emotional schema therapy is entirely on verbal exchanges; bodily aspects such as patient-therapist synchrony do not receive much consideration. In light of the empirical material examined in this chapter, which demonstrates the crucial role synchrony can play in establishing rapport and emotion regulation, one might suspect that this omission is unfortunate.

Only a few researchers have highlighted the importance of nonverbal dynamics at the therapist-patient system level, but we might expect a surge of interest in light of a recent systematic account of adequate methodologies for synchrony in psychotherapy. In a series of promising studies of psychotherapy sessions, Ramseyer and Tschacher (2011, 2014, 2016) showed bodily synchronizations between patient and therapist, involving whole-body- and head-movement-coupled patterns. Using the recently developed *motion energy analysis*, which automatically codes body movements from video footage, the authors analyzed over one hundred videotaped sessions with outpatients suffering from nonpsychotic, mainly affective disorders. Importantly, the studies showed that higher nonverbal synchrony was associated with higher symptom reduction, as well as positive correlation between psychotherapy outcome, synchrony, and the quality of therapeutic alliance. These results support earlier findings concluding both that the quality of the therapeutic alliance is an accurate predictor of the outcome of the therapy (Horvath et al. 2011; Flückiger et al. 2012) and that the quality of bodily interaction predicts psychotherapeutic success (Rasting and Beutel 2005; Tschacher and Pfammatter 2016). However, the results also have indicated that distinct mechanisms operate in head and body synchrony, which correlate differently with overall and session-level therapy success.

Building on these findings, Koole and Tschacher (2016) combined results from studies on therapeutic alliance and synchrony, integrating them into what they refer to as the *interpersonal synchrony (in-sync) model*. According to this model, "synchrony facilitates the alliance, which in turn promotes the patient's emotion-regulatory skills" (Koole and Tschacher 2016, 12; see also Tschacher, Rees, and Ramseyer 2014; Veenstra, Schneider, and Koole 2017).

The basic proposal is worth quoting at length:

> The core idea of the In-Sync model is that the alliance emerges from the coupling of the neural activity of the brains of the patient and therapist. The more tightly patient and therapist's brains are coupled, the better the alliance. Of course, patient and therapist's brains do not communicate directly. Their coupling can thus be achieved only indirectly, through the mutual coordination of the patient and therapist's behavior and experiences. This mutual coordination is achieved through synchronous activities of the patient and therapist. Synchrony thus helps to establish the alliance, which in turn promotes adaptive emotion regulation in the patient, and thereby good therapeutic outcomes. (Koole and Tschacher 2016, 7)

This account explains several results and provides an explanation of why nonverbal interactional synchrony in psychotherapy is correlated with the quality of the alliance and with therapy outcomes. The basic idea is that (1) bodily synchrony enables interbrain coupling, which, in turn, (2) creates an alliance between patient and caretaker. Finally, (3) the alliance supports adaptive emotion regulation in the patient. But the in-sync model also raises several questions. The following sections will explore some of these questions and argue that thinking in terms of intersomatically scaffolded emotion regulation provides a better explanation.

7.5 Three Questions

Looking closer at point (1), we may agree that taking neural activation into account offers a more comprehensive understanding of how synchrony involves neural and non-neural parts of the body. However, one may argue that—at least in light of currently available material—it is artificial to insert a divide and posit a causal relationship between interbodily and interbrain couplings. Instead, there are several alternative options.

First, it might be that interbrain synchrony is the neural *counterpart* of the observed interbodily synchrony, in which case one would simply be describing neural and non-neural parts of the same phenomenon. Second, the agent's bodily and neural activity could *together* form one half of the dyadic system, with the other agent's bodily and neural activity forming the other half. In both cases, it is not clear why one should prefer an explanation of the underlying mechanism exclusively in terms of inter-brain coupling. The question is why one would eliminate or relegate inter-bodily synchrony from an explanation.

Taking a closer look at point (2), it is worth digging deeper into the claim concerning the relationship between synchrony and therapeutic alliance. Although it rings true that synchrony helps establish some kind of positive relationship, it would be misleading to posit a unidirectional causal relation between synchrony and alliance. In a sense, we can say that synchrony supports a positive relation and sense of attachment, which then leads toward more synchronization, and so on. But again, separating the smooth and dynamic flow into static fragments seems like the less attractive option. If, however, we think in terms of active scaffolding, then it is no longer puzzling that synchrony can be seen both as causally implicated in and as constituting a part of the positive relationship.

Moving to point (3), it is not entirely correct to say that the alliance supports adaptive emotion regulation in the patient—in particular, not if we take alliance to refer to a kind of cooperation based on some shared understanding of a task. Rather, it is more precise to say that synchrony characterizes alliance and that it underlies emotion regulation. Saying that the alliance itself drives emotion regulation would be less informative and accurate, just like saying that in the case of couples, it is love and respect that drives the relevant regulatory processes. Instead, it is more accurate to maintain that synchrony is the bodily correlate or marker of a productive alliance, while also constituting part of the intersomatic scaffold that drives emotion regulation.

Finally, it is important to mention that these considerations are merely intended to point to possibilities for empirical investigations into intersomatically scaffolded processes. Whereas the overall impression is that emotion regulation can be actively scaffolded onto forms of synchronic bodily interaction, we currently lack bottom-up studies executing interventions on synchrony to test the effects on emotion regulation.

7.6 Synchrony and the Praecox Feeling

Operationalized systems have not been efficient in identifying patients with early signs of schizophrenia, but experienced clinicians are often said to be able to "intuitively" detect fine-grained aspects for a preonset diagnosis. The idea of recognizing the specific indications in an intuitive way goes back to the Dutch psychiatrist H. C. Rümke, who used the term *praecox feeling* to describe the psychiatrist's experience of an inability to establish contact with the patient. Such a feeling is described as a kind of bodily unease emanating in the interviewer that reflects the detachment and alienation of the patient (Grube 2006; Ungvari et al. 1997; Peters 2000). More precisely, Rümke emphasized three characteristics (Varga 2013). First, he maintained that the praecox feeling arises from a lack of smooth, prereflective interpersonal exchange. Second, he noted that it emerges in the absence of verbal indicators and concluded that the praecox feeling is linked to the nonverbal motor behavior of the patient. Third, the feeling is described as the psychiatrist's "inability to empathize" with the patient: "The investigator examining a sufferer from schizophrenia notices something out of the order within himself" (Rümke 1941, 336).

Rümke (1941) argued that the praecox feeling of the psychiatrist should be an important guiding factor in diagnosis, but many have questioned its diagnostic reliability. In a study investigating this issue, experienced psychiatrists interviewed sixty-seven patients previously unknown to them, with symptoms indicating schizophrenia spectrum disorder (Grube 2006). The accuracy of the psychiatrists' intuitive judgments was very high. The percentage of correct assignment for schizophrenia diagnosis was 86.6 (according to ICD-10) or 83.6 (according to DSM-IV). The psychiatrists reported the presence of the praecox feeling at very early stages of the encounter, before commencing a more comprehensive examination. Moreover, the strength of the praecox feeling correlated with the severity of cognitive and emotional disturbances, distorted self-perception, and compromised communication.

These findings raise questions about the underlying mechanisms that enable these intuitive judgments, but standard approaches have difficulty offering suitable answers. The most straightforward answer seems to be that experienced psychiatrists have developed a fine-tuned perceptional apparatus that allows them to detect fine-grained bodily signs in the patient, below the radar of conscious awareness, on which their judgments are based. However, though this may surely be a part of the picture, it alone does not explain why the perceived signs lead to a specific feeling of unease. A more developed explanation starts with the suggestion that experienced psychiatrists have developed an increased ability to (a) engage and synchronize with patients and that (b) synchronization is reduced or lacking in interactions with patients. While (a) can be seen as an uncontroversial assumption that is consistent with standard accounts of expertise, (b) is well-supported by empirical material. Individuals with prodromal syndromes, as well as unmedicated recent-onset patients, exhibit significantly reduced nonverbal behavior (including facial expression, body and head movement, gaze, and gesture; see Kupper et al. 2010; Dimic et al. 2010; Brüne et al. 2008; Walther et al. 2015), and the intensity of their symptoms is closely associated with their level of nonverbal behavior (Lavelle, Healey, and McCabe 2013). A recent study using the computer-based motion energy analysis has confirmed these findings and showed that reduced synchrony is associated with negative symptoms (Kupper et al. 2015). For this reason, Kupper et al. (2015, 2) suggest that "the subtle lack of synchrony in verbal and non-verbal interactive behaviors may be the key to clinicians' intuitive experience of a 'praecox feeling' owing to the presence of schizophrenia in patients."

Although one may agree with this conclusion, which is further supported by decreased nonverbal synchrony being associated with lower rapport (Bernieri and Rosenthal 1991; Lavelle, Healey, and McCabe 2013), the explanation cannot be merely that the psychiatrist somehow senses the patient's inability to synchronize. This would not explain why the praecox feeling is experienced by the psychiatrist as there being "something out of order" within herself. To accommodate this, we may say that experienced psychiatrists have developed an increased ability not only to (a) engage and synchronize with patients, but also to (c) skillfully compensate for desynchronizations, which would provide the basis for a good rapport and eventually also for scaffolded emotion regulation. This is not a contentious claim; it has been shown that healthy individuals interacting with schizophrenic patients exhibit increased gesturing (Lavelle, Healey, and McCabe 2013), which can be understood as a form of unconscious compensation for the lack of synchrony. Although this is an assumption that would require empirical support, it seems safe to assume that an experienced psychiatrist would be much more skilled at employing forms of compensation than individuals with no background in healthcare. Moreover, in the course of their clinical experience and supervision, we may suppose that experienced psychiatrists have acquired (d) a heightened awareness of their own bodily changes. When we consider these points together, they provide support for the following view: (d) makes it possible for the expert psychiatrist to experience the "unease," which is the experiential signature of the psychiatrist's failure to achieve both (a) and (c). The experiencing of (a) alone would not explain the psychiatrist's experience of there being "something out of order" within herself.

To be clear, the explanation on offer here is in no way complete. Nonetheless, it provides impulses toward a theoretical reorientation and further research, while being consistent with both Rümke's conclusions and recent empirical material on nonverbal behavior in schizophrenia. A more comprehensive picture requires identifying the bodily signals psychiatrists single out and explaining why this occurs exclusively in schizophrenia, as other mental disorders, such as depression, also exhibit diminished bodily expressivity and motor behavior. Although further studies are needed to investigate the connection between the patient's condition and the psychiatrists' experience, the previous sections have offered compelling reasons for taking seriously the concept of synchrony in such investigations.

7.7 Conclusion

Having explored how cognition can be scaffolded onto the sensorimotor system, this chapter expanded the scope of the inquiry and addressed a particular form of simple scaffolding, in which the scaffold includes not only the agent's sensorimotor apparatus but also external structures that adaptively guide behavior. In such cases, the vehicles of cognition individuate externally, constituting a neural, bodily, and extrabodily assembly.

The bulk of the chapter was dedicated to cases of intersomatic scaffolded cognition and the role they can play in mental disorders. The chapter traced the development of interactive skills that develop in childhood and make possible particular synchronic interaction patterns in early infant-caretaker interactions. Such synchronic interactions enable and sometimes drive certain cognitive processes in children and adults. Consistent with the evidence presented throughout this book about scaffolding, offloading, and integration, it is likely that synchrony originally developed to sustain group membership and bonding but was later repurposed to be used in interpersonal emotion regulation.

Unquestionably, improving our comprehension of synchrony has important implications for theory, practice, research, and training. Moreover, there is reason to think that interactional synchrony is a valid operational construct for investigating diagnostic aspects, as well as therapeutic success. A recent study evaluating the effectiveness of the manualized body-oriented psychotherapy for schizophrenia assessed negative symptoms and interactional synchrony before and after therapy. The study found a large inverse relationship between interactional synchrony and negative symptoms (Galbusera, Finn, and Fuchs 2016). However, though body-oriented psychotherapy intervention appears to ameliorate negative symptoms and interactional synchrony, the mechanisms through which this occurs remain unclear. Further complicating the picture is that body-oriented psychotherapy encompasses several heterogeneous techniques (for a review, see Röhricht 2009).

8 Scaffolded Social Cognition

8.1 Social Cognition, Mindreading, and Autism Spectrum Disorder

By shining light on the coupling of sensorimotor systems in dyadic interactions, chapter 7 brought into focus some of the features that make us into uniquely social beings. There are of course many nonhuman animal species that have developed complex forms of social life that involve communication, the division of labor, and the maintenance of social hierarchies. There is, however, a sense in which human beings display uniquely complex social skills that enable and support the use of tools and multifaceted representational systems, which then further enhance our ability to comprehend our fellow human beings. Questions about the nature and genesis of social cognition have motivated philosophical inquiry for centuries, but these efforts were significantly invigorated by scientific theories aiming to explain its development in terms evolutionary pressures (Humphrey 1976; Whiten 2000). While living in groups has imparted on the individual adaptive advantages, it has also increased the range of social interactions, competition, and forms of reciprocal dependence. The combination of these aspects generated particular selective pressures to develop enhanced reasoning capacities, which could be deployed, for instance, to identify deception and nontrustworthy individuals. From such a background, it is often maintained that a specific mechanism developed to ensure adaptive responses to the changed social environment (Cosmides and Tooby 1992; see also Sterelny 2003, ch. 10).

Some argue that this is a domain-specific mechanism. This view draws support from studies showing that independent of our beliefs, we automatically interpret another agent's bodily movements as intentional actions, anthropomorphize objects and creatures, and offer intentional descriptions of very simple moving triangles. Moreover, many have assumed that this view is supported by studies on autism spectrum disorder (ASD), which involves impaired mindreading but leaves other, nonsocial skills unharmed. As we shall see over the course of this chapter, the picture is much more complicated.

The perhaps most significant question that captivated researchers concerns the nature of the mechanisms and cognitive processes utilized in mindreading. There are two aspects in the course of research that seem particularly important in light of our goals. First, though social cognition need not involve mental states, research has traditionally focused on the skill to infer a mental state in the mind of another and thus on the comprehension of behavior in terms of causally powerful mental states that enable us to predict, explain, and manipulate behavior, leading to fitness-enhancing forms of communication and coordination (Heyes and Frith 2014; Krebs and Dawkins 1984).[1] This capacity, commonly referred to as *mindreading*,[2] minimally requires representing a mental state with its content (a proposition specifiable by "that" clauses) and comprehending the relations among the mental state, behavior, and environmental circumstances (Hutto, Herschbach, and Southgate 2011). Mindreading admits of degrees, with children being able to grasp a narrower variety of mental states than adults.

Second, to shed light on the relevant mechanisms, researchers were particularly interested in impairments to social cognition that characterize several mental disorders.[3] It has been argued that our strong propensity to attribute mental states—even to inanimate objects—may be an important factor that renders social cognition vulnerable to dysfunction (Brüne and Brüne-Cohrs 2006). But regardless of whether this is correct, impairments to social cognition occur in numerous mental disorders, like ASD and schizophrenia, but also depression (Wang et al. 2008) and bipolar disorder (Kerr, Dunbar, and Bentall 2003; Wolf, Brüne, and Assion 2010). Researchers have argued that several characteristic symptoms can be explained by underlying deficits in mindreading (e.g., Baron-Cohen 1997, 2008; Frith 2004). It is therefore not surprising that ASD, the mental disorder in which difficulties in social cognition are perhaps most prominent, has attracted a significant amount of attention. Since the emergence of ASD as a distinct clinical entity in the 1940s, researchers have stressed that though children with

ASD exhibit continued interest in engaging with objects, often in repetitive ways, they do not show much interest in social interactions. Today, it is accepted that ASD has an important genetic component, and the (behavioral) diagnostic criteria include not only deficits in communication and interaction but also a restricted range of activities, interests, and imagination (Happé 1994).

Nonetheless, mindreading remains at the center of current debates on ASD. Mindreading, spontaneous attention to people, and joint attention deficits appear early on and are considered characteristic of the whole spectrum. An influential approach that by and large follows cognitivist principles explains several diverse social impairments in ASD in terms of deficiencies in "theory of mind." The idea is that individuals with ASD lack a proper "theory" of mental states, the deployment of which allows individuals to infer a mental state in the mind of another (Baron-Cohen 1997). Some researchers continue to consider "mindblindness" as the principal feature of ASD, whereas others now favor multicausal explanations and propose that ASD is in fact a compound of cognitive impairments, with individuals being differently affected in the domains of mindreading, central coherence, and executive function (Joseph, Tager-Flusberg, and Lord 2002; Pellicano et al. 2006). Although many now agree that no single approach will be able to explain the entire range of social cognition problems in ASD, different theoretical approaches to social cognition offer distinctive perspectives on how to understand impaired processes in individuals with ASD.

This chapter will provide an investigation of social cognition in ASD. It will be argued that emerging findings on sensorimotor difficulties indicate that ASD may suitably be considered in light of ASC. Rather than explaining the impairments exclusively in terms of mindreading deficits, the aim is to show that highlighting the scaffolded nature of social cognition may offer a new understanding of the symptomatology.

8.2 A Tacit Theory

The idea that social understanding is supported by a "theory" is familiar in philosophy,[4] but the term "theory of mind" was coined by Premack and Woodruff (1978) in their study of primates' capacities to understand their conspecifics. Developmental psychology utilized the term to designate the ontogenetic expansion of progressively more sophisticated theories, and it now refers broadly to processes that enable mindreading (Goldman 2006, 10). The view that everyday mindreading is underpinned by some kind of "theory" that explains behavior in terms of attributed mental states reflects a broadly cognitivist line of reasoning that many consider appealing for several reasons. First, just as one can argue that understanding grammatical sentences in a language requires representing the grammar of that language, one can argue that understanding others and predicting their behavior requires a resource that links mental states to behavior and that is somehow represented, perhaps in a language-like structure or some connectionist network. Second, adopting a realist stance, one could argue that this theory is enormously successful at offering correct predictions of behavior because it accurately tracks actual mental states and their associated properties (see, e.g., Fodor 1987). Third, the view that social understanding is based on an empirical theory is often taken as the best available explanation. As Carruthers (2015, 502) notes, "We have no idea how to explain the causal processes involved except by appealing to something amounting to a tacit theory." Fourth, though some have complained that the idea of deploying a theory is too far removed from the phenomenology of social understanding, the claim is that theorizing is subpersonal.[5]

Some hold that the theory is innate and triggered by experiences at a particular developmental stage (e.g., Carruthers 1996; Leslie 1987, 1988, 1992; Baron-Cohen 1997; Fodor 1983; Scholl and Leslie 1999). It develops from a module that is informationally insulated and outside of conscious control. The claim is that theory of mind is widespread, invariant throughout history, and acquired via the same developmental pathway that indicates species-level genetic inheritance. Otherwise it is hard to explain how young children acquire a broad knowledge of human psychology in environments often described as characterized by a poverty of stimulus. Others hold that the theory is gradually acquired and draw parallels to how scientists develop and deploy scientific theories (Gopnik and Wellman 1994; Gopnik, Meltzoff, and Bryant 1997) or scientific models (Godfrey-Smith 2005). The idea is not that folk psychology involves law-like sentences, but that children, much like scientists, respond to anomalous observations, initially with additional ad hoc conjectures, then with the rejection of the established theory and the acceptance of a new theory in light of the evidence.

Whether the theory is innate or acquired, the overall approach to mindreading is consistent with a general cognitivist framework. It subscribes to the epistemological separability thesis, maintaining that social cognition can be investigated without having to attend to the body. Moreover, "theorizing" does not require a specific body, such that Martians or computers mastering the relevant regularities could in principle form proper predictions and explanations (see, e.g., Churchland and Churchland 1990).

8.3 Toward Two-Level Accounts

To test when mindreading emerges, researchers have explored the acquisition of the concept "belief" by investigating when children begin attributing false beliefs to others. In a series of influential false-belief task experiments, Wimmer and Perner (1983) demonstrated that children younger than three years do not attribute false beliefs to others. The rough design of the experiment is this: Person A places an object in Box A and leaves the scene. Person B then enters and switches the location of the object from Box A to Box B. Person B leaves and Person A returns. The prediction is that a participant with a functional theory of mind would attribute to A a false belief, which leads A to look in Box A. Children with a dysfunctional theory of mind, or younger than four years, would say Box B, which is standardly interpreted to show that they fail to master the concept of belief, ignorant of the possibility that beliefs can misrepresent reality.

Employing Wimmer and Perner's (1983) false-belief task, Baron-Cohen, Leslie, and Frith (1985) tested three groups of children, with Down syndrome, with ASD, and typically developing children. The study demonstrated that though most children with ASD were unsuccessful in mindreading tasks and at understanding false beliefs and mentalistic stories, they were able to comprehend stories that described physical events. Because children with Down syndrome were able to pass the test, the conclusion was that children with ASD do not exhibit general cognitive shortcomings (their mean IQ was significantly higher than children with Down syndrome) but have impairments in a domain-specific capacity. This has led some researchers to maintain that there is a particular, probably modular mechanism for generating metarepresentations.[6] Leslie (1988, 1994) argued that the domain-specific "theory of mind mechanism" uses a specific representational system and forms the basis for mindreading. Moreover, the emergence of pretend play in children offers early evidence of such a mechanism, which later enables the child to form metarepresentations—that is, mental representations of other people's mental representations. Whereas a decoupler mechanism used in pretend play underlies the ability to form mental metarepresentations, in ASD the mechanism is damaged, which explains both why children with ASD fail to engage in pretend play and why they fail to acquire full-fledged mindreading. The idea was that "not being able to understand that another person 'pretends' had the same cause as not being able to understand that another person 'intends,' 'knows,' or 'believes'" (Frith 2012, 2078).

While questions remained about whether the children lack the relevant concepts or are merely insufficiently skilled in applying them, others resisted the idea that much of everyday social understanding and interaction requires mindreading (Gallagher 2001a, 2001b, 2008). Based on pioneering work by Shaun Gallagher, interaction theory illustrates how social cognition can occur through online embodied social interaction (e.g., Reddy 2008; Fuchs and De Jaegher 2009).[7] But the greatest challenge that has led to questions about what the standard false-belief task is actually measuring comes from a number of studies showing that significantly younger children can in some sense attribute false beliefs.

Several studies have demonstrated that much younger children are able to pass nonstandard false-belief tasks. Studies using a "violation-of-expectations" protocol showed that fifteen-month-old infants exhibit behavior that indicates attributing a false belief to an agent (Onishi and Baillargeon 2005; see also Baillargeon, Scott, and He 2010). Similar results were obtained in thirteen-month-old infants (Surian, Caldi, and Sperber 2007). Fourteen-month-old infants are surprised when a person looking for an object neglects a deceptive cue (Song and Baillargeon 2008), while seventeen-month-olds more clearly track the agent's false belief (Southgate, Chevallier, and Csibra 2010). Eighteen-month-olds demonstrate comprehension of the goals of others and exhibit willingness to assist in achieving them (Warneken and Tomasello 2006), and twenty-five-month-olds can anticipate actions, even though this requires attributing a false belief (Southgate, Senju, and Csibra 2007). Moreover, certain instances of social cognition like understanding the intentions of others are present not only in preverbal stages in children, but also in orangutans (Call and Tomasello 1998) and rhesus monkeys (Santos, Nissen, and Ferrugia 2006).

Impairments in social cognition in ASD show a more nuanced picture. Some children with ASD displaying high verbal ability engage in symbolic pretend play, even if their pretend behavior is frequently stereotyped and not spontaneous (Lewis and Boucher 1988; Whyte and Owens 1989; Jarrold et al. 1994). This is mostly taken to indicate the presence of metarepresentational capacities. In addition, a large percentage of individuals with ASD manage to pass false-belief tasks (Happé 1995; Iao and Leekam 2014; Scheeren et al. 2013). Children with higher verbal ability do not experience profound difficulties (Frith 2012; Senju 2012), which is consistent with the general view that performance on theory of mind tasks is directly related to linguistic ability (Sharrock and Coulter 2004). Children and adolescents with high-functioning ASD understand sophisticated mentalistic stories (involving emotions and false beliefs, sarcasm, and bluff) at a level that is not significantly different from that in normally developing children. Also, adults with high-functioning ASD can acquire mental concepts and infer the relevant mental states, although they do not implicitly track false-belief states of others (Schneider et al. 2013) and do not master the ability to smoothly deploy their skills in social interactions (Scheeren et al. 2013). Numerous studies found that participants with ASD who were able to pass standard false-belief tasks nonetheless lacked the kind of spontaneous anticipatory gaze that indicates some form of mindreading (Senju et al. 2009; Senju et al. 2010).

These findings point to the necessity of a more multifaceted approach to mindreading, in which the capacity to pass standard false-belief tasks and metarepresentational abilities only provide a partial picture of the processes and mechanisms involved in social cognition. They open the way toward two-system accounts, which could accommodate findings in studies using both standard and nonstandard tests, as well as the possibility that standard false-belief tasks detect mindreading deficits that, at least in part, are manifestations of a more fundamental impairment in social cognition. Two-system accounts differentiate between an implicit system that is activated early on and a more explicit system that develops later in life. The former is fast and inflexible, whereas the latter is cognitively more demanding, but flexible, and allows complex reasoning in terms of mental concepts (Apperly and Butterfill 2009; Apperly 2011; Southgate, Senju, and Csibra 2007). The terminology in the literature tends to vary, but the general idea is this:

Implicit mindreading: A type of social understanding that develops at an early stage and involves an implicit understanding of mentality, even if the mindreader is incapable of attributing psychological states to others. The criterion is merely that the mindreader's behavior is systematically dependent upon changes in the psychological states of others.

Explicit mindreading: A type of social understanding that develops at a later stage and involves the capacity to attribute psychological states to others. The criterion is that the mindreader's behavior is systematically dependent on representations of the psychological states of others.

In light of this distinction, the seemingly conflicting findings can be understood as resulting from the fact that the studies generating them actually investigate different systems. The studies using implicit measures to investigate gaze patterns and anticipatory looking offer a window into the workings of implicit mindreading, whereas studies using standard false-belief tasks shed light on the later-developing explicit mindreading. The distinction not only helps make sense of the empirical material but also advances the debate on the nature of the impairments in ASD. For instance, it enables making sense of how it is possible to pass standard false-belief tasks while lacking the kind of spontaneous anticipatory gaze that indicates implicit mindreading. Some influential researchers now maintain that though individuals with ASD can acquire explicit mindreading by means of compensatory learning, the deficit in the early-appearing, implicit mindreading is crucial for understanding a wide variety of social difficulties (Frith 2012).

8.4 Simulation

A view that often is presented as an alternative account of mindreading proposes that mindreading involves generating internal simulations of others' mental states (Gordon 1986; Heal 1996). While some of the early work highlights processes accessible to consciousness and subject to some degree of voluntary control, it is now commonly held that simulation may be conscious or subpersonal and understood in the replication or duplication sense, not in the computational-modeling sense (Goldman 2006, 36). Proponents of this view typically see the role assigned by cognitivism to folk psychology as unjustified and oppose the cognitivist view that mindreading necessarily involves representations in propositional format (Gallese 2007, 2011, 2014). As Gallese et al. (2009, 107) put it, "What forms the basis of our capacity (and the capacity of nonhuman primates) to understand the intentions behind others' behavior is a more basic functional mechanism that exploits the intrinsic functional organization of the motor system." So rather than having to consult a theory, we simply mentally "step into the shoes" of the other and use our psychological mechanisms to run offline simulations, which is more cost-effective than theory-driven mindreading because it does not require storing and accessing knowledge about human psychology.

Somewhat similar to the two-system accounts, the simulationist view now distinguishes two levels of mindreading (Goldman 2006):

Low-level mindreading: A "comparatively simple, primitive, automatic" mirroring process for the comprehension of simple motor plans, sensations, and emotions. It occurs "largely below the level of consciousness" (Goldman 2006, 113) and involves an "unmediated resonance," which is enabled by the fact that performing an action and observing it activate shared cortical networks.

High-level mindreading: A process for the comprehension of more complex mental states such as propositional attitudes, with some components of the process accessible to consciousness and subject to voluntary control (Goldman 2006, 147). A process of "enactment imagination" generates pretend mental states and runs an offline decision-making process on them.

Some features of high-level mindreading raise questions. To start a successful simulation process, one must reduce complexity (Dennett 1987). There are potentially infinite aspects and combinations of aspects that could be relevant to simulate, which highlights the need for a selection process that can only occur in light of knowledge about the simulated entity that is organized into a theory-like complex. But in that case, simulation seems to collapse into theorizing. In reply, one might grant that theory-involving simulation is required when simulating an essentially different target organism, but not if the simulating organism comprehensively resembles the target (Goldman 1989, 2006). In the latter case, assuming that the simulation process is comparable to the process that drives the behavior of the target, simulation may indeed succeed in producing the right state. However, unless the process is automatic, even in this case some initial "knowledge" about the relevant similarity is required. This might not amount to a "theory" involving a set of folk-psychological laws, but it could perhaps be described as something like an algorithm for starting simulations.

Low-level mindreading appears much less susceptible to this objection. Gallese and Goldman (1998) early on linked low-level mindreading to the *mirror neuron system* (Goldman 2012), which is constituted by several brain regions that activate when an individual performs a particular action and when she sees another person performing that same action. Mirror neuron function is understood in terms of direct matching, such that perceived actions in others are directly mapped onto the motor scheme of the perceiver (Rizzolatti and Sinigaglia 2010). Single neuron studies demonstrated mirror neurons in macaque monkeys (Gallese et al. 1996; Fogassi et al. 2005). But there are also indications of the presence of mirror neurons in the human brain (Mukamel et al. 2010). The mirror activation of mirror neuron populations in motor cortices during action observation was confirmed by other methods, including fMRI (Buccino et al. 2001; Oosterhof et al. 2010; Kilner et al. 2009) and electroencephalography (Cochin, Barthelemy, and Roux 1999), although it should be mentioned that imaging and neurophysiological studies only offer indirect evidence.

There are important differences in the function of the mirror mechanism in humans and nonhuman primates. First, in humans, mirror neurons are also present in the supplementary motor area and the medial temporal lobe, which is predominantly active in memory tasks (Gallese, Rochat, and Berchio 2013). More recent studies suggest that the network in humans is broader than hitherto assumed, involving the somatosensory and premotor cortex (Keysers, Kaas, and Gazzola 2010; Caspers et al. 2010). Second, the activation patterns in human agents display a higher level of abstraction. For instance, the recognition of motor intentions is modulated by contextual information, and it is readily extracted from a large variety of movements involving different physical properties (Cattaneo et al. 2009; Cattaneo et al. 2011). There are some indications that motor areas can have a predictive function. In some cases, when enough contextual cues are accessible, they can become active before observing another's action (Kilner et al. 2004). Third, though mirror neurons are assumed to enable understanding motor intentions in both human and nonhuman primates, it is also maintained that they have taken on further functions in humans—for instance, assisting in language development (Rizzolatti and Arbib 1998).

Given this link between low-level mindreading and the mirror neuron system, is low-level mindreading susceptible to the objection of collapsing into theorizing? Simulationists argue that low-level mindreading is not vulnerable to this objection because the relevant neural mirroring activity is "unmediated." To illustrate the unmediated nature of mirroring, Goldman invokes studies showing overlapping neural activation while both feeling pain and perceiving pain (Singer et al. 2004), feeling being touched and seeing another person being touched (Keysers et al. 2004), and smelling repulsive odors and observing faces that express disgust (Wicker at al. 2003).

However, though these studies do demonstrate mirroring, they use unambiguous settings. Other studies using less unambiguous settings show that mirroring, at least in the case of action, is in fact mediated by the context. In an fMRI study by Iacoboni et al. (2005), participants observed a grasping hand without information about the context of the grasping, with context only (scenes containing a teapot, a mug, cookies, a jar), and with a grasping hand in two different contexts, suggesting that the intention related to the grasping was "drinking" or "cleaning." Although both conditions involved graspable objects, the activation patterns were different. Such findings cannot be explained merely by invoking the characteristic mirroring properties. Instead, the authors infer the activity of "logically related" neuronal activity, maintaining that "there are neurons that are visually triggered by a given motor act (e.g., grasping observation), but discharge during the execution not of the same motor act, but of another act, functionally related to the observed act (e.g., bringing to the mouth)" (Iacoboni et al. 2005, 0533). But if it is correct that mirroring offers a context-sensitive understanding of the intentions of others, then it is unclear in what sense simulation activity could be "unmediated." There is no claim that low-level simulation collapses into theorizing on the basis of folk-psychological laws, but simulationists need to account for this type of context sensitivity.

8.5 Autism Spectrum Disorder and Simulation

Concerns about collapse notwithstanding, simulationists offer a detailed approach to understanding mindreading difficulties in ASD and have addressed some issues raised against the simulationist view. First, it has been argued that neither the conventional conception of simulation as involving pretense nor simulation as involving matching qualifies as a suitable model for mirror neuron activation (Gallagher 2007). Opting for a better alternative, Gallese (2014) characterizes simulation in terms of *reuse*: the neural circuits of the motor system that primarily program and execute action are reused to assist social cognition. Second, the relationship between high- and low-level mindreading is further clarified, maintaining that high-level mindreading is "preceded, both from a philogenetic and ontogenetic point of view, by non-propositional forms of social understanding" (Uithol and Gallese 2015, 4). This view dovetails with evidence that the mirror network is something we share with primates, while allowing that, in humans, low-level mindreading is not only antecedent to high-level mindreading but also coexists with it. Thus, the direct mapping of the perceived action into one's own motor repertoire that enables low-level social understanding may remain active when human agents deploy high-level mindreading.

The simulationist hypothesis offers a particular take on social cognition in ASD, sometimes referred to as the *broken mirror theory*. Contrary to what cognitivism would predict, simulationists maintain that global impairments in the functioning of mirror neuron networks may underlie problems with social cognition in ASD. Variants of this thesis advance slightly different claims (see Hamilton 2013). Some maintain that the impaired simulation system in the mirror neuron network causes difficulties with mindreading and language use in ASD (Iacoboni and Dapretto 2006; Oberman and Ramachandran 2007). As Oberman and Ramachandran (2007, 322) suggest, a "developmental impairment of functional simulators, especially the MNS, may be the unifying mechanism that underlies the deficits in imitation, ToM, empathy, and pragmatic language." Others hold that the breakdown of elementary self-other mapping in ASD leads to problems with both imitation and high-level mindreading (Williams et al. 2001; Williams et al. 2006; Gallese 2006; Gallese et al. 2009). In this view, high-level simulation develops on a foundation provided by an undamaged self-other mapping function, implemented by mirror neuron activity. Disrupted mirror neuron activity then could explain low-level mindreading problems, stereotyped mimicking (echolalia), difficulties in copying actions, and limited ability to generate the kind of "offline" imitation involved in high-level mindreading. For instance, Williams (2008, 84) argues that "the mechanisms that underpin imitation and self-other matching are fundamental to our understanding of autism."

The evidence for this hypothesis is, however, far from robust, and there is considerable controversy about the extent to which it is specific for the symptomatology of ASD. A recent systematic review examines twenty-five neuroscience studies investigating the activity of the mirror neuron system in ASD (Hamilton 2013). Hamilton maintains that weakly localized measures of mirror neuron function like electroencephalogram (EEG) and TMS yield mixed results and do not offer evidence for abnormal activity in ASD because they cannot rule out the impact of top-down factors on the mirror neuron system. Findings from localized measures (fMRI), which measure the BOLD signal in mirror neuron regions, offer a clearer test of the hypothesis, suggesting that emotional stimuli lead to anomalous responses (Dapretto et al. 2006; Grèzes et al. 2009). Nevertheless, the fact that these differences do not occur with nonemotional action stimuli significantly weakens the hypothesis.

Overall, available data on mirror neuron abnormalities in ASD is not robust enough to support the idea of a global mirror system dysfunction (Hamilton 2013; Keysers, Thioux, and Gazzola 2013). Some argue that the data suggesting mirror system dysfunction could be explained by atypical attention to social stimuli (Enticott et al. 2013) and as downstream neuropsychological consequences of a poor developmental history of social engagement and learning (Vivanti and Rogers 2011; Colombi, Vivanti, and Rogers 2011). It is important to add, however, that the lack of a global mirror system dysfunction does not undercut the more modest interpretation that the mirror neuron system has a central role in social cognition and its activity is altered in ASD.

8.6 Scaffolding, Disintegration, and Autism Spectrum Disorder

The broader view that this chapter aims to support is consistent with this modest interpretation, but it focuses on sensorimotor deficits in ASD based on the ASC framework and its forms of scaffolding introduced in previous chapters. Some researchers early on pointed to characteristic sensorimotor disturbances in children with ASD (Damasio and Maurer 1978; Vilensky, Damasio, and Maurer 1981),[8] maintaining that these might have a key role in understanding the condition (Rogers and Pennington 1991; Leary and Hill 1996). Whereas sensorimotor disturbances remain under-researched compared to social, cognitive, and communicative aspects (Mosconi and Sweeney 2015), they are reliable early indications of ASD (Grossberg and Seidman 2006).[9] The impairments include motor control (gross and fine), vestibular control, oculomotor functions, and skeletomotor functioning (for reviews, see Gowen and Hamilton 2013; Mosconi and Sweeney 2015; Fournier et al. 2010; Eigsti 2013; Dowd, Rinehart, and McGinley 2010; Bhat, Landa, and Galloway 2011). Studies have shown reduced postural stability, balance, and coordination (Mari et al. 2003); difficulties with coordinated arm movements, reaching, and grasping (Glazebrook et al. 2009); problems controlling force and direction when throwing (Staples and Reid 2010); difficulties with fine motor movements (Freitag et al. 2007); and problems with anticipating timing consequences of actions and postural adjustments (Cattaneo et al. 2007; Rinehart et al. 2001). Based on kinematic studies, other researchers show impaired use of perception-action coupling (Whyatt and Craig 2013a, 2013b). Importantly, motor skills in infants with ASD strongly predict later social skills (Sutera et al. 2007; Piek and Dyck 2004) and can be used to detect deficiencies at an early stage, when they are not concealed by mechanisms that have been developed to compensate (Mostofsky et al. 2006).

In the remainder of this chapter, the literature on sensorimotor impairment in ASD will serve as a basis for putting forward a specific hypothesis according to which low-level mindreading does not merely involve neural mirroring and reuse. Consistent with ASC's commitment to epistemological inseparability, the hypothesis is that the reused mechanism not only consists of the mirror system, but makes part of a scaffold that also involves the body, its movements, muscle activations, and sensory feedback. Thus, intact social cognition requires not just an intact functional configuration of the cortical motor system, but also the operation of simple scaffoldings that offer peripheral sensory feedback. But if this is true, then we may expect that the *disintegration* of parts of this scaffolding will lead to difficulties in social cognition. If low-level or implicit mindreading reuses a system constituted by neural components and bodily components, then the lack of appropriate muscle activation (e.g., in automatic mimicry), action kinematics, and sensory feedback may impede the automatic imitation that is involved in social cognition.

8.6.1 Simple Scaffolding and Disintegration

In chapter 3, the term *simple scaffolding* was introduced, along with the somewhat surprising finding that Botox injections can produce improvement in certain depressive symptoms. Typical emotional states in depression involve the contraction of the corrugator muscle, which, when temporarily paralyzed, interrupts responses to negative stimuli. Due to the ways in which mental processes are scaffolded onto sensorimotor circuits, manipulating the body may causally affect relevant emotional and cognitive processes. Could a similar way of thinking help explain some of the important symptoms in ASD?

We may start with studies on emotional contagion, which offer support for the view that automatic mimicry is lacking or significantly weakened in ASD. Mimicry is typically considered to explain emotional contagion, which is taken to indicate bodily attunement with others (Hatfield, Cacioppo, and Rapson 1994). Individuals with ASD exhibit very few signs of contagious yawning, while diagnostic severity is correlated with susceptibility to contagious yawning. The conclusions we can draw from such studies are limited, however, as mimicry often involves very subtle matching patterns of facial muscle activity in the observer, which remains undetectable for unaided observation and can only be measured with electromyography. Such muscle-activation patterns in reaction to others' expressions and movements constitute a perceptually driven automatic mimicry, which can be present at the age of five months (Isomura and Nakano 2016) and the magnitude of which is greater in the case of observing expressions of emotion (Moody and McIntosh 2011). This mimicry serves as a simple scaffold in social cognition (Moody and McIntosh 2006), with which the obstructed facial muscle responses interfere (Oberman, Winkielman, and Ramachandran 2007).

Individuals with ASD have decreased access to this resource. For instance, though individuals with high-functioning ASD are capable of intentionally producing typical facial emotional expressions, when measured with electromyography they exhibit significantly reduced amounts of automatic facial mimicry (McIntosh 1996). In a more recent study, the facial muscle activity of participants was monitored while they observed pictures of faces with happy and angry expressions. Compared to age, intelligence, and gender controls, adults and adolescents with ASD capable of voluntary mimicry did not exhibit automatic mimicry of facial expressions (McIntosh et al. 2006). Perceiving facial expressions of happiness and anger led to nonspecific muscle-activity patterns in adults (McIntosh et al. 2006) or no discernable responses (Beall et al. 2008). In experimental conditions explicitly requiring emotion recognition, children with ASD exhibited normal, but delayed levels of response (Oberman, Winkielman, and Ramachandran 2009).[10]

Emotion recognition appears to be affected by the *disintegration of the bodily scaffolding* (i.e., muscle activation, movement, etc.), and similar issues characterize the kinematics of action execution and the automatic mimicry and understanding of actions. A study using motion-tracking technology investigated the movement kinematics (velocity, acceleration, and jerk) of simple arm-waving movements and found that individuals with ASD display irregular arm movements, the magnitude of which is significantly correlated with symptom severity (Cook, Blakemore, and Press 2013). The study also found evidence of a direct relationship between kinematics and action perception. In an independent task on biological motion perception, participants observed a moving hand with minimum jerk motion (characteristic of the movements of human beings) and constant velocity (characteristic of the movements of robots). Participants were instructed to label the movement as natural or unnatural. The results showed that the more atypical the participant's own kinematics, the less likely they were to label movements displaying typical human kinematics as natural, which, conceived in terms of interventions, indicates that the participants' own kinematic profile in some way actively scaffolds their perception of movements displaying typical kinematics.

Atypical kinematic profiles not only confirm problems with motor control but also are associated with an altered perception of actions. A study using electromyography investigated the link between action execution and automatic imitation during action and perception and recorded the activity of the mylohyoid muscle, which reinforces the mouth floor and assists jaw-opening movements (Cattaneo et al. 2007). The participating children, a group with ASD and a control group consisting of typically developing children, were tested in action-execution and action-observation conditions. In the first condition, they were asked to pick up food and eat it or pick up a piece of paper and put it in a container. In the second condition, the participants observed a child executing one of these two tasks. The study found different muscle activity in children with ASD, and the dissimilarities during action execution corresponded to the dissimilarities during action observation. Typically developing children in the execution condition showed increasing anticipatory muscle activity before and during grasping, and their muscle activity reached its greatest peak when they began opening their mouths. In contrast, children with ASD exhibited no anticipatory muscle activity during grasping until the food reached the mouth. In the observation condition, typically developing children exhibited muscle activity when observing somebody reaching and grasping the food, whereas children with ASD did not show such activation.

Whereas these studies indicate the lack of appropriate muscle activation and action kinematics, others emphasize increased *sensorimotor noise* (sensed change that is not interpreted as a signal) and re-entrant sensory feedback (Mostofsky et al. 2006) that also contributes to motor coordination. Torres et al. (2013) argue that corrupted and "noisy" afferent peripheral information can result in irregular movements, as well as inconstant reafferent kinesthetic proprioception, which can obstruct the coordination and anticipatory control of movements. As Brincker and Torres (2013, 1) put it "many behavioral-level findings can be re-defined as downstream effects of how developing nervous systems attempt to cope and adapt to the challenges of having various noisy, unpredictable, and unreliable peripheral inputs."

Overall, the mirroring activity involved in successful low-level social cognition is not merely neuronal; it also involves the integration of sensorimotor resources into a scaffolding that supports social cognition. But in that case, exploring sensorimotor problems in ASD might shed light on some aspects of the impairments in social cognition. Decreased muscle-activation patterns, atypical action kinematics, and increased sensorimotor noise are factors that can lead to atypical responses, to the failure to precisely anticipate the consequences of one's own movements, and to impaired automatic mimicry and mapping between one's own motions and those of others. When these impairments reach a certain threshold, it is reasonable to assume cascading downstream implications. Perceptually driven, bottom-up automatic mimicry is then not available as a reliable feedback resource.

8.6.2 Intersomatic Scaffolding

The last sections were mainly concerned with simple, intrasomatic scaffolding that involves the sensorimotor system and offers a supporting structure that cognitive processes utilize. But we may expect that the disintegration of this scaffolding will lead to difficulties in integrating external structures to adaptively guide behavior. More precisely, we may predict difficulties with creating the kind of intersomatic scaffolding explored in chapter 7, in which the vehicles of certain cognitive processes include the body of another human being in synchronic interactions. Can these predictions be supported?

Recall that originally, behavioral synchrony probably developed to sustain group membership, but was later repurposed and integrated into the machinery of dyadic emotion regulation. As we have seen, decreased mimicry weakens the sense of connectedness between individuals and increases perceived psychological distance, which also reduces mimicry. From this background, it appears uncontroversial to assume that decreased automatic mimicry will lead to decreased levels of alignment in the kind of rhythm and timing coordination that dyadic emotion regulation depends on. Thus, because synchronous interaction in caretaker-infant dyads takes on the role of supporting the emotion regulation of the child, decreased synchronization may be expected to lead to decreased intersomatic emotion regulation.

Most studies examining social cognition and motor function have concentrated on mimicry and imitation, but interpersonal behavioral synchrony, which represents another aspect of interpersonal coordination, suffers from comparable impairments in ASD. Fitzpatrick et al. (2016) used pendulum coordination tasks to assess synchrony in adolescents. Adolescents with ASD demonstrated both less spontaneous synchronization and less intentional synchronization when the explicit goal involved coordinating with another person. Based on earlier sections of this chapter, it appears defensible to assume that sensorimotor problems in part explain synchronization difficulties. But it is important to note that the disruptions are selective and most tangible in social interactions. In a different paradigm, individuals with ASD asked to synchronize their movements with a dot on a computer screen performed normally (Koehne et al. 2016). Consequently, it is specifically social attention that is associated with impaired intentional synchronization. However, deficits appear to extend to quasi-social spontaneous synchronization. In a study using a rocking chair paradigm, parents were instructed to adjust their movements to a certain frequency that the children could not hear. The study found that typically developing preschoolers exhibit spontaneous tendency to rock in a symmetrical state with their parents, whereas children with ASD do not (Marsh et al. 2013).[11]

In a sequence of studies, Fitzpatrick et al. (2013, 2016) found that school-age children and adolescents with ASD had weaker synchronization capacities than controls. Another study investigated the relationships between synchronization ability and motor control problems (Fitzpatrick et al. 2017). On three synchronization tasks (synchrony battery, imitation battery, and interpersonal hand-clapping), children with ASD performed motor movements that were slower and less consistent in both spacing and timing over course of the interaction. The relatively poor motor skills alone do not fully explain the findings. For instance, the difference between the performances of the groups was most significant for behavioral synchrony tasks requiring children to match the movements of the experimenter.

Given that children with ASD often exhibit motor problems and rigid behavioral patterns, it is likely that basic forms of dyadic regulation of emotional states would be more challenging. It may be expected that the tight parent-child exchanges that constitute behavioral synchrony and underlie dyadic emotion regulation operate differently than with typically developing children. A recent study by Guo et al. (2017) investigated dyadic emotion regulation in mother-child interactions videotaped in the home environment. Compared to dyads with typically developing children, dyads with children with ASD displayed different dyadic emotion regulation, with different patterns of alterations of arousal aimed at maintaining an optimal emotional state. Compared to dyads with typically developing children, dyads with children with ASD had fewer mutual positive interactions, spent overall more time and longer intervals engaged exclusively with objects, and spent more time in mismatched states. Typically developing children stayed longer in mutual positive engagement states. The analysis of microcoded play video footage of a child and mother during play at home reveals increased amounts of negative arousal and less dyadic synchrony than in dyads with typically developing children. These interactions offer less support for emotion regulation processes, even though mothers of children with ASD are usually highly skilled interactors attuned to the needs of their children. Whereas mothers of typically developing children decrease the use of active strategies like physical redirection and prompting, mothers of toddlers with ASD continue to use these strategies, along with vocal comforting strategies. Still, even so, and even though children with ASD seek their mothers' assistance when trying to deal with distress as much as their typically developing peers (Hirschler-Guttenberg et al. 2015), dyads with a child with ASD spend more time in inflexible states and have reduced coregulation capacity (Larkin et al. 2015).

Most of these studies focus on mother-infant interactions, but the same pattern is found in studies investigating the behavioral coordination of father-infant dyads. Trevarthen and Daniel (2005) conducted a frame-by-frame microanalysis study of monozygotic twin girls based on home videos made by the parents. The girls were eleven months old, and one of them was diagnosed with ASD before she reached the age of two. The girl later diagnosed with ASD showed very little eye contact, and her interactions with the father exhibited less attention and anticipation of emotional changes. The lack of synchrony with the parent hindered the father's attempts to support her motor activities and to scaffold her emotion regulation.

It is worth pausing to accentuate that these conclusions are based on a relatively low number of available studies. They receive some support from other findings, showing, for example, that interventions targeting joint engagement have a positive effect on dyadic emotion regulation (see, e.g., Gulsrud, Jahromi, and Kasari 2010). Future studies will have to evaluate to what extent factors other than basic motor deficits—for instance, in more complex rhythmic timing—may lead to synchronization differences. But the findings indicate that the difficulties in behavioral synchronization over time contribute to decreased dyadic emotion regulation in the child. Given that dyadic emotion regulation plays a role in both the real-time regulation of the child's emotional state and in the development of adaptive emotion regulation strategies, early synchronization dysfunctions likely contribute to the social skills deficit in ASD, making it even more difficult to build social and emotional bonds and respond to social cues.

8.7 Conclusion

Cognitive approaches have long dominated ASD research, placing emphasis on theory-of-mind deficits, even though the onset of sensorimotor abnormalities precedes social and communicative deficits. Some researchers early on lamented that research in ASD largely disregards the body and movement and describes social cognition in purely mentalistic terms (see, e.g., Leary and Hill 1996). Since then, there has been an increasing attention to sensorimotor processes and their role in interaction. Rather than explaining the impairments exclusively in terms of mindreading deficits, the aim of this chapter was to integrate recent empirical work with the conceptual distinctions of the ASC framework and to show that a better comprehension of the scaffolded nature of social cognition may offer new perspectives on the symptomatology.

Focusing on low-level mindreading, it was argued that individuals with ASD exhibit disintegration in the scaffolded social cognitive mechanism that includes neural and non-neural components. The claim is, of course, not that the sensorimotor problems will explain all or most features of the behavior of individuals with ASD. The view defended in this chapter is consistent with indications that early sensorimotor problems lead to atypical movement, perception, prediction, and sensory feedback integration, which are crucial for social cognition. Moreover, the view is also consistent with the heterogeneous behavioral symptomology in ASD.

The scaffolding assisting social cognition can become disintegrated at many different sections and at many different developmental stages, with each combination leading to markedly different downstream consequences. More importantly, the findings drawn upon to support the thesis indicate that shifting attention from mental symptoms to offering a fine-grained account of sensorimotor changes may lead to identifying specific markers for early diagnosis. Such sensorimotor markers could inform diagnostic subtypes and lead to conceptualizing intervention. The advantage would be that sensorimotor markers can be measured objectively, do not depend on language ability, and could be used to assess individuals with mild to severe symptoms. This, however, requires a more thorough grasp of the specificity of the sensorimotor problems in ASD compared to other conditions. Atypical movements and kinematic patterns occur in developmentally delayed individuals and in various conditions like Parkinson's disease, cerebral palsy, and ADHD, without instigating the kind of difficulties in social cognition that characterize individuals with ASD.

As to novel possibilities for intervention, it is warranted to assume that increasing automatic mimicry would enhance social cognition in individuals with ASD. However, given its automatic nature, it is difficult to conceive of ways to directly target it. Therefore, further research is required to determine whether improving instructed interpersonal coordination and synchronization has the potential to increase the automatic propensity for mimicry and improve social skills in children with ASD.

Concluding Remarks

The impetus for this book arose from discerning several intriguing changes in the contemporary philosophical landscape. A broad methodological reorientation is characterized by a gradually decreasing emphasis on a priori methods. Instead, philosophers increasingly conceptualize the philosophical study of the mind as continuous with empirical work and progressively draw on cognitive science for scientifically informed philosophical reflection. The growing engagement in such *empirically informed philosophy of mind* has contributed to the emergence of a field of study that offers fertile interfaces between philosophy and cognitive science, allowing for the application of philosophical resources to the subject matter of scientific inquiries and their calibration in light of empirical findings.

This broad methodological reorientation has played a key role in the emergence of productive fields. Much of the philosophical work in this book was propelled by two distinct shifts within empirically informed philosophy of mind that invite philosophical engagement with the theoretical commitments of emerging conceptual frameworks and research programs. The first shift can be described in terms of an intensified study of the *embodied mind*. The theoretical basis of cognitive science is slowly moving away from cognitivism, which views mental processes as abstract formal processes, or as activation patterns in neural networks that exclusively are realized in the brain and can be comprehended in abstraction from the body and the environment. In contrast, EC subscribes to the idea that various features of cognition profoundly depend upon features of the agent's body and environment, such that these sometimes play non-negligible causal or constitutive roles in her cognitive life.

The second shift can be described in terms of an intensified study of the *disordered mind*, expressing an acknowledgment of the convergence of the explanatory and therapeutic concerns of psychiatry and interdisciplinary inquiries into the mind. Somewhat akin to manipulations in experimental studies, exploring features of mental disorders can help discern the nature of the mechanisms that underlie normal cognitive functioning. An empirically informed reflection on psychopathological phenomena constitutes a valuable resource for testing theories, offering real-life cases instead of hypothetical ones that feature in philosophical thought experiments.

Combining these two shifts offers the possibility of complementing contributions and distinctive insights into cognitive processes and of exploring potential benefits for the understanding and treatment of mental disorders. Traditionally neglected bodily aspects of psychopathology can now be systematically addressed, and experimental manipulations in controlled environments deployed to unearth organizational, structural, and functional features of cognition can be combined fruitfully with studying "naturally" occurring changes in individuals with mental disorders in ecologically valid environments. Overall, the *epistemological conjecture*, which emphasizes the noteworthy potentials for philosophical research and clinical practice, proposes that a full understanding of our cognitive lives requires considering its dependence upon features of the (non-neural) body and environment and its vulnerability to disintegration.

Following a relatively common approach in empirically informed philosophical inquiries, this book offered a theoretical contribution, followed by application to and calibration by empirical research in mental disorders. The first steps involved examining central theses of traditional cognitivism, disentangling differing positions and theoretical commitments in current debates and introducing several fine-grained distinctions within the ASC framework. These steps were essential to be able to offer a more precise focus in a relatively nonhomogenous area of research. But neither ASC nor the overall pluralist outlook suggested in chapter 4 claim to be able to solve or bypass all discordances in different accounts. Instead, the book advocates that cognitive processes are too complex to be explained in a single, unified framework, yet without abandoning the effort to amalgamate relatively dissimilar scaffolds of the mind into a loosely knit framework.

ASC adopts the epistemological inseparability thesis and utilizes the term *scaffolding* to attain the suitable amount of conceptual flexibility required for the task. The term denotes artificial or nonartificial supporting structures, but also the *active* process of structuring, while remaining neutral on the boundaries scaffolds can have, as well as on their physical implementation (including physical, mental, or hybrid entities and processes). Without claiming to provide a complete taxonomy of forms of scaffolding (leaving out, for instance, forms of environmental and cultural scaffolding), the second part of the book sought to apply and adjust the conceptual and theoretical resources of ASC. It was demonstrated that the forms of scaffolding introduced in the first part of the book—together or separately—can offer useful resources for comprehending prominent features of mental disorders and for providing new ideas for therapeutic measures. The focus on the various ways in which concepts and cognitive activities are scaffolded onto the sensorimotor system offers fertile perspectives on particular symptoms in depression and ASD that have remained elusive.

It is important to acknowledge some limitations. The position proposed and defended in this book is crucially informed by and dependent on a rich variety of empirical findings, without being able to do justice to the exponentially growing material that would be worth considering. At various junctures in this book, several important questions left unanswered were underscored, often accompanied by rough suggestions, which are hopefully sufficiently well-formed to be of use for further research. Moreover, one important limitation arises from the range of mental disorders under discussion. The DSM-5 (2013) contains more than two hundred diagnostic categories, and in light of their complexity and sheer variety, it would be misguided to think that any single volume can provide a comprehensive account that would warrant many generalized statements. The range of the conditions considered in this book is limited to a few common disorders, which severely restricts permissible generalized conclusions but does not hinder demonstrating the usefulness of the approach for understanding particular conditions.

It is also critical to point out that even if one finds ASC convincing, the arguments in the book do not support the view, embraced by quite a few scholars, that the biological brain is not an adequate unit for cognitive scientific investigation or not the most suitable target for psychiatric intervention. Instead, the point is merely that sometimes the most adequate unit for cognitive scientific investigation is the hybrid ensemble of neural, bodily, and environmental elements. But ASC is skeptical of the view sometimes suggested by researchers in cognitive neuropsychiatry that we will soon be able to uncover the cognitive mechanisms that are responsible for psychiatric disorders such that we will be able to reduce psychiatric disorders to neural disorders. This idea has a long history in psychiatry. From its beginnings as a medical discipline, there were two competing types of explanation in psychiatry: a biological approach pursuing a causal explanation in terms of bodily states, and an interpretative approach pursuing comprehension by creating meaningful links between mental events (Phillips 2004). By the nineteenth century, psychiatrists maintained that mental disorders are best understood as "brain diseases" and defended a view of psychiatry as an emerging academic medical discipline that investigates various forms of neuropathology (see Varga 2015, ch. 2). It is not difficult to comprehend the lasting appeal of this idea. Psychiatry often has been criticized as unscientific because it lacks a diagnostic classification system based on biological criteria. The hope often is that if mental disorders could be reductively explained, then malfunctioning cognitive mechanisms could be identified with a neural malfunction, and psychiatry could finally share the same scientific status as other areas of medicine.

This hope to reduce psychiatric disorders to neural disorders, however, assumes that the activation of neural mechanisms is not only necessary but also sufficient for a cognitive process to occur (Sprevak 2011). The ASC framework presented in this book opposes this view, although it does not doubt that impaired brain mechanisms are heavily involved in mental disorders. In the ASC view, reducing mental disorders to neural disorders is impossible because the cognitive mechanisms implicated in mental disorders are often actively scaffolded in ways that do not respect the brain-world boundary. But if the mechanisms can at least in part involve non-neural entities and processes, then they will defy reduction even if researchers utilize the most advanced imaging technologies. As this book has shown, identifying the relevant cognitive mechanism and its impairments in a mental disorder often requires describing the activity of the scaffolding that involves sensorimotor domains and environmental elements, which often are structured by human agents.

As a last-ditch effort, cognitivists could split the overall mechanism into neural and non-neural parts and maintain that only the neural part deserves the label *cognitive*. But this would come at the price of an unnecessarily complicated account with reduced explanatory power. Also, it would perhaps misleadingly suggest that therapeutic intervention should always aim at restoring "normal" neural functioning. Although ASC concurs that such interventions are often the most appropriate ones, in many cases successful intervention may target one or multiple non-neural parts of the scaffolding. Especially in those cases in which ASC offers a better, integrated explanation of clusters of seemingly separate symptoms in particular mental disorders, we have reasons to think that a non-neural component of the hybrid cognitive ensemble may in fact constitute a suitable target for psychiatric intervention. Thinking further along these lines will not only stimulate the interaction between cognitive scientists and clinicians, but also illustrate how much the reshaping of our physical and social environment also means reshaping our cognitive capacities.

Notes

1 Philosophy, Psychiatry, and Cognitive Science

1. As van Gelder (1998b, 134) rightly notes, in many cases "it would be wrong to think of the philosopher of cognitive science as a person always quite distinct and separate from the psychologist, linguist, computer scientist, etc., and as engaged in completely different and independent tasks."

2. As an example of how philosophical ideas directly motivate scientific investigations in psychology, Dennett's work on intentional action generated fruitful research on children's judgments about false beliefs (Dennett 2009; Thagard 2009).

3. Alternatively, philosophy *of* cognitive science also can be pursued by using the tools of ethical and sociopolitical philosophy—addressing, for instance, the effects of scientific knowledge about the mind and reflecting on its place in the context of liberal democracies.

4. For instance, Marr's (1982) influential investigation of the visual system focused on explicating in information-processing terms the algorithms by which an information-processing task is solved.

5. The study of cognitive psychopathology now increasingly is taking on a role that resembles that of brain lesion studies in cognitive psychology.

6. For some examples, see Fuchs and Schlimme 2009; Drayson 2009; and Maiese 2016.

2 Cognitivism, Clinical Cognitivism, and Embodied Cognition

1. It is hard to deny that computationalism is a part of a comprehensive account of cognitive psychology. But the question whether computation is sufficient for cognition is not straightforwardly settled, unless perhaps one also maintains that consciousness is reducible to computation. However, if one thinks that computation is not sufficient for phenomenal consciousness and that cognition involves phenomenal consciousness, then computation is not sufficient for cognition.

2. There are several naturalistic theories explaining how mental items can be said to represent such as Fodor's asymmetric dependence theory (1990, 1994) and Millikan's teleosemantics (1984, 1989, 2004). In Fodor's (1990) causal approach, mental representation is characterized by asymmetric dependence. Agent A's thoughts about x are meaningful in virtue of a causal relation between a mental representation of x and some part of the world that is x. A's thoughts of x are about x because it is in some sense a law of nature that x causes the mental representations of x. Fodor (1990) takes into account that a non-x can cause the mental representations of x, but argues that this is asymmetrically dependent upon the fact that x causes mental representations of x (for a discussion, see Kriegel 2014). For example, "smoke" tokens can be caused by fire, smoke, cigarettes, and so on. The point is that cigarettes and fire would not cause "smoke" tokens if smoke did not. As Fodor (1990, 93) puts it, "'(s)moke' tokens carry information about fire (when they're caused by smoke that's caused by fire). But they don't mean fire because their dependence on fire is asymmetrically dependent on their dependence on smoke." The theory generated some counterexamples (see, e.g., Adams 2000), but Fodor later clarified that he only aimed to offer sufficient conditions. In contrast, Millikan's teleosemantics explains content not in terms of its causes, but in terms a naturalistic etiological notion of function (as selected effects), which derives its normativity from natural selection. Millikan draws on research on complicated signaling systems in animals like honeybees and beavers. Bees perform a complex "dance" that signals to other bees the location of nectar. Details of the dance systematically inform bees about the direction to the nectar, the distance, and its attractiveness. The dance is representational (representing direction, distance, and attractiveness) but exhibited in creatures the traits of which can be explained in evolutionary terms. At the same time, representations in bee dance admit of misrepresentation. In this sense, bee and human cognition are comparable, though the former operates on a simpler level, blending cognitive and motivational aspects and deploying "pushmi-pullyu representations" (Millikan 2004). Very roughly put, content derives from proper function, which resists definition in causal terms. Although SNAKE can be tokened in the absence of snakes, what determines the representational content of SNAKE is what *should* be responsible for causally producing it, which is determined by the direct proper function of the producing mechanism.

3. To understand some of the motivation, it is useful to recall that support for classicism came in part from the device known as the "Turing machine," developed to formalize the notion of computation. Alan Turing used a mathematical notion of computation as an algorithmic process and demonstrated that an abstract Turing machine could function as a general machine capable of evaluating any computable function. The machine consists of an endless tape, divided into squares, with a reading device that can read the content of each square. The head can move the tape in both directions, and, using binary notation, it is able to erase and write on a square. Importantly, such a simple machine operating with a very limited set of

rules could complete a wide range of computations and was capable of the kind of problem-solving that we usually think requires a human mind (MacDonald and MacDonald 1995). The classical Turing machine became the reference model for computation because it was able to simulate other computational models and resembled the emerging digital computer, and Turing's notion of computation was applied to describe brain function (McCulloch and Pitts 1943).

4. For instance, Churchland (1995, 322) notes that "it is now evident that the basic unit of memory is the synaptic weight configuration."

5. Compulsions were hypothesized to arise from suboptimal learning in neural network models (see also Caspar, Rothenfluh, and Segal 1992).

6. In the following, the term *cognitive therapy* is used to refer to several closely related approaches that build on cognitivism. Many have adopted the term *cognitive-behavioral therapy* (CBT), in part because most researchers in the field had a behavioral background (Ingram, Siegle, and Steidtmann 2009, 76). It can be slightly misleading, however, because the emphasis is explicitly on cognition as the appropriate target of treatment. In the last decades, new approaches that all fall under the general umbrella of CBT have mushroomed, including acceptance commitment therapy, mindfulness, and dialectical and radical behavior therapy.

7. Clinical cognitivism is computational only in a broader sense, without much emphasis on fine processing details or computer modeling, but the main idea of cognition as information processing has had a very significant influence.

8. For instance, Varela, Thompson, and Rosch (1991, 9) oppose the notion that cognition is essentially the manipulation of representations. On the other hand, Clark (1997) remains sympathetic to the idea of internal representations (but aspires to rethink their formats and contents), whereas Lakoff and Johnson (1999) do not essentially oppose computationalism.

9. The ties to Merleau-Ponty here are weaker, as Merleau-Ponty does not explicitly discusses metaphors in his work (in fact, he deals relatively little with language in general). There are, however, some aspects that indicate clear parallels, particularly in chapter 6 of *Phenomenology of Perception*, in which Merleau-Ponty describes language as essentially an embodied activity, portraying words as the linguistic counterparts of bodily gestures and referring to the body's ability to convert "a certain motor essence into vocal form" (Merleau-Ponty [1962] 2005, 211).

10. The phenomenological approach has had a significant influence on psychopathology, offering rich analyses of subjectivity, temporality, and embodiment in mental disorders (for some recent examples, see Sass and Parnas 2017; and Fuchs 2005).

3 The Scaffolded Mind

1. This is not an exhaustive list, and there are, of course, several other open questions—for instance, about the continuity of life and mind or about the extent to which the entanglement of body and mind is mediated through representations of the body (see Goldman 2014, 93).

2. *Engineered* should be understood in a broad sense, not only as exclusively involving consciously planned activities. Also, not all scaffoldings discussed in this book are engineered in any substantial sense.

3. This is why several accounts will not be considered, even though they meet the first two selection criteria. These include, for instance, ecological psychology (Richardson et al. 2008), AI (Brooks 1999), and dynamical systems theory (van Gelder and Port 1995; Erlhagen and Schöner 2002; Kelso 1995).

4. See Colombetti and Krueger's (2015) productive use of Sterelny's work to understand emotions.

5. Mainly located in the ventromedial prefrontal cortex.

6. There is a debate about whether it is the absolute intensity of the signal or the contrast between the signals that matters (see Guillaume et al. 2009).

7. The notion of manipulation is clarified in detail in chapter 5.

8. Some have argued that a sudden separation of these individuals from their home environment equals disconnecting them from parts of their minds (Dennett 1996; Drayson and Clark 2013).

9. Debates surrounding Clark's work have led to "second-wave" approaches to motivate the hypothesis, which subscribes to the *complementarity principle*, maintaining that the hybrid integration of neural and environmental vehicles characterizes human cognition (Menary 2010a, 235; see also Menary 2007; Rowlands 2010; Sutton 2010). External entities are complementary tools (rather than mereological parts) that are "integrated" (Menary 2010a, 227) or "amalgamated" (Rowlands 2010, 83) into a cognitive whole, without having to be functionally equivalent.

10. Taking this idea even further, some maintain that the internet can function as an external memory, forming a part of distributed cognitive systems (Sparrow, Liu, and Wegner 2011).

4 ASC, Ontological Inseparability, and a Pluralist Landscape

1. There are also "internal" tensions that require attention, however. For instance, we have seen that though they all support the epistemological inseparability thesis, only some of them also endorse ontological inseparability.

2. Some argue that a mark could prevent the debate from becoming stalled (Rowlands 2010) and would promote convergence between rival conceptions of cognition (Buckner 2015). Some go even further and argue that "without an independently motivated account of the What of cognition, quarreling about its Where is pointless" (Walter and Kästner 2012, 18).

3. I will assume a particular reconstruction of the debate, as proposed by Clark (2008b), Wheeler (2010), Rowlands (2010), and Sprevak (2009).

4. Some emphasize fine-grained features of human memory (e.g., Miller's rule, generation effect) that extended systems lack (Rupert 2009), while others argue that cognitive processes are exclusively realized by particular lower-level (neuronal) processes (Adams and Aizawa 2008; Fodor 2009). The point is that fine-grained characteristics "should guide us in determining what memory is like and what really differentiates cognitive processes and mechanisms from non-cognitive processes and mechanisms" (Adams and Aizawa 2010, 10).

5. Some caution is warranted because some sentences, like "the reason plants turn their leaves toward the light is to maximize photosynthesis" (Adams and Garrison 2013, 347), seemingly involve a reasons-based explanation that does not underwrite cognitive processes.

6. Some have objected that Rowlands's criteria are too narrow and include only biological systems (Adams and Garrison 2013, 343–344). However, there is nothing in Rowlands's account that would necessitate such a limitation, which becomes clear when he discusses the deployment of knots by knivu officers (Rowlands 2010, 124–125). In that case, the knots are conceived as having a proper function, but not in a strictly biological sense.

7. Rowlands draws a parallel to the activity of identifying and laying bricks and argues that there are certain parts of this process that the agent has epistemic authority over. The agent can identify individual bricks, recognize the characteristics of good mortar, determine the optimal amount of cement to use, and so on. The agent has "epistemic authority over how each brick is laid" and is "epistemically responsible for certain parts of the process and product" (Rowlands 2010, 153).

8. Opponents of the extended mind hypothesis (EMH) do not necessarily commit to the view that certain fine-grained characteristics studied by cognitive scientists (e.g., Miller's rule) are necessary for something to count as short-term memory. Nonetheless, the point often seems to be that "findings such as this should guide us in determining what memory is like and what really differentiates cognitive processes and mechanisms from non-cognitive processes and mechanisms" (Adams and Aizawa 2010, 10, 70). Of course, much will depend on what *guide* means here, and some opponents might not accept this reconstruction of the debate. But for the purposes of this chapter, I accept Clark's and Wheeler's point, according to which EMH opponents hold that certain fine-grained features (nonderived representations

governed by idiosyncratic processes studied by cognitive scientists) distinguish cognitive from noncognitive processes.

9. Moreover, Adams and Aizawa (2008, 2010) and Fodor (2009) argue that the substrates that realize cognition operate with "non-derived representations" that exclusively occur in brain-bound systems (Adams and Aizawa 2010). Interestingly, Adams and Aizawa (2008) claim that it is only a contingent matter of fact that cognition occurs in the brain. But it is worthwhile to note that embracing fine-grained or coarse-grained functional similarity does not necessarily determine one's position in this debate. On the one hand, while Otto's extended memory is coarse-grained, some future device may allow him to display the same kind of fine-grained functional characteristics that ordinary human memory does. On the other hand, opponents of EMH could, in principle, accept that some humans without parts necessary for the usual fine-grained characteristics can nonetheless be regarded as remembering. Thanks to an anonymous referee for pointing this out to me.

10. Such fine-grained features of human memory also include the *generation effect*, the finding that information is better recalled if it is self-generated rather than achieved by other means, such as by reading.

11. Wheeler argues that the fact that both commonsense and scientific concepts of memory are broad enough to cover such cases supports a *locationally uncommitted* account of cognition. Wheeler's (2010, 259) more precise claim is that *not all* cognitive traits require specification at a coarse level of grain.

12. Although this point will not be further pursued, it is worth mentioning that, at least ideally, relying on *folk* psychological intuitions requires independent arguments. This is because it is well-known that scientific-psychological findings have provided numerous facts about the workings of the mind that cannot be accommodated by strong folk intuitions. But the claim is not that the fine-grained characteristics that cognitive scientists study do not matter or that cognitive science could never correct our folk psychology about what constitutes a cognitive process.

13. Particularly relevant ideas are voiced in Lewis's (1980) "Mad Pain and Martian Pain" and in Block's (1991) criticism of "chauvinist" (psycho)functionalism.

14. Or perhaps "mad Martian pain," as in Lewis's (1980) thought experiment.

15. Such *cognizer-first* attribution of cognition to Martians becomes more plausible if we accept the (more or less overtly) humanoid nature of imaginary Martians. See this point in section 2.

16. Rupert (2013, 11) has contested some of the inferences based on the Martian thought experiment, arguing that "we have the intuition that Martians have memories, but only because we can imagine that it would be useful to use the term 'memory,' under at least some circumstances, were we to come across such Martians

and study them scientifically." I shall take a different approach and investigate to what extent the Martian thought experiment produces valid results.

17. It is disputed whether thought experimenting is best seen as some kind of scientific method (generating empirical knowledge) or an armchair method (generating a priori knowledge). Philosophers with a naturalistic bent think that intuitions offer empirical access to the world, providing "shortcuts" in empirical investigation that condense prior analytic experience (DeWitt 2006). Most accept that thought experiments need not restrict themselves to cases that could plausibly occur. So long as a scenario is conceivable, such as the Martian scenario, it is legitimate to use it to elicit intuitions.

18. For instance, East Asian intuitions on various well-known philosophical thought experiments are significantly different from intuitions that Western philosophical literature assumes as general (Weinberg, Nichols, and Stich 2001). Moreover, intuitions are also influenced by educational background and gender (Swain, Alexander, and Weinberg 2008; Stich and Buchwalter 2011). The Martian thought experiment is less susceptible to criticism in this regard, which is why I focus on internal validity.

19. For instance, in the case of thought experiments in ethics, some worry that the ability to form "correct" moral intuitions may not include "outlandish" cases (Elster 2011).

20. It should be noted that the position defended here does not entail embracing Wilkes's point that thought experiments ought to be informed by current scientific knowledge.

21. In 1897, K. Lasswitz's novel *Auf Zwei Planeten* introduced the figure of the humanoid, yet in many ways superior, Martian that still occupies a prominent place in popular culture. They can interbreed with humans, but are cognitively, technologically, and ethically superior to us. Hugo Gernsback's *Ralph 124C 41+* has humanoid Martians living among humans on Earth, while A. N. Tolstoy's *Aelita, Queen of Mars* from 1923 uses a Martian setting to explore life in a class-based society. The tradition of conceiving of Martians as humanoid continues in novels like Eric Frank Russell's *Men, Martians and Machines* (1955) and Ray Bradbury's *The Martian Chronicles* (1950). Philip K. Dick's *Martian Time-Slip* (1964) describes a slowly dying humanoid Martian race (the "Bleekmen"). In television and movies, CBS's *My Favorite Martian* (1963–1966) features a humanoid Martian who crashes on Earth, whereas *The Martian Chronicles* (1980) depicts humanoid Martians with golden skin. *Mission to Mars* (2000) describes Martians as tall, peaceful humanoids, whereas humanoid red Martians play a major role in Disney's science-fiction movie *John Carter* (2012). Surely, there are also many examples of less humanoid descriptions of Martians. But it is difficult to dispute that the concept of the Martian as humanoid

has a prominent position in popular imagination. When we are asked to imagine Martians, it is therefore at least likely that we imagine them as humanoid.

22. It should be noted that parts of the argumentative strategy in this chapter might appear odd: one could object that throughout this chapter, I simultaneously rely on and criticize thought experiments. This is why it is important to clarify that I do not advocate the wholesale rejection of thought experiments. The point is merely that some of them suffer from underdetermination and lead therefore to questionable results.

23. A general objection to calibration was originally raised by Cummins (1998).

24. Again, EMH critics may not agree with this reconstruction of the debate. Although they do seem to think that EMH's view of location and grain-parameter are at odds with common sense, EMH will emphasize that they draw on scientific findings and not on intuitions about grain. The idea is that fine-grained processes operating with nonderived representations typically studied by cognitive scientists distinguish cognitive from noncognitive processes. I remain skeptical about this argument but provide in this context no arguments of my own. Instead, as mentioned in the introduction, for the purposes of this chapter, I adopt Clark's reconstruction of the debate.

25. One also could argue that clashing intuitions might be seen as a symptom of something else—namely, that the intuitions elicited are themselves systematically biased in some sense. Some have argued that the mere presence of conflicting intuitions in trained philosophers can be taken as prima facie evidence for theory-ladenness. As Schroer and Schroer (2013, 1276) note, "In the case of observation, defenders of the theory-ladenness often point to ambiguous figures—e.g. the duck/rabbit—as concrete examples where theory seems to affect observation. In the case of intuitions, we submit that clashes of intuitions among professional philosophers can play a similar role—i.e. they are prima facie evidence for the existence of theory-laden intuitions." If it is right that clashes of intuitions may indicate the existence of theory-laden intuitions, then this might just be enough to transfer the burden of proof to the thought experimenter. If it turns out that the relevant intuition is theory-laden, its ability to function as theory-neutral evidence for a specific philosophical theory is limited. A theory-laden intuition can support a theory only if the theory with which the intuition is laden has different informational content than the theory supported by that intuition.

26. Rosch and Mervis (1975, 575) explain that "a family resemblance relationship consists of a set of items of the form AB, BC, CD, DE. That is, each item has at least one, and probably several, elements in common with one or more items, but no, or few, elements are common to all items."

27. The approach proved successful. For instance, when extrapolated to the domain of emotion, prototype theory generated correct predictions (Fehr and Russell 1984; for an approach consistent with this idea, see LeDoux 1996, 16).

28. The former hold that vehicles of cognition are symbols with certain formal properties, while the latter think that they are patterns of node activation.

29. For arguments in favor of pluralism in chemistry, see Slater 2005 and Dupré 1993.

30. Buckner (2015) offers an interesting account that has not received consideration here. Propelled by issues arising in the demarcation of cognition from association, Buckner suggests thinking of "cognition" as a "relaxed" natural kind, in line with Boyd's (1991) work on homeostatic property clusters. Compared to other approaches to the mark of the cognitive, conceptualizing cognition in terms of clustered properties is of course closer in spirit to what is suggested in this chapter. The clustered properties are inclusive, relatively loosely defined, and reverse-engineered from scientific practice. Nonetheless, cognition is "identified by eight specific types of representational flexibility that enable agents to pass typical tests for cognition" (Buckner 2015, 309), which excludes instances of nonrepresentational and associative learning from the realm of the cognitive.

5 Scaffolding and Dependence

1. It is worth noting that some operate with an even broader concept, where scaffolding merely denotes the "means through which other structures or competencies are constructed or acquired by individuals or organizations" (Wimsatt and Griesemer 2007), which covers both large organizations (schools) and small biological entities (cells).

2. A general shift to explanatory considerations, some of which have productively engaged recent work on mechanistic explanation, has been advocated by Sprevak (2010).

3. This also holds in cases in which interventions are physically impossible (Woodward 2003, 71).

4. As Woodward (2003, 30) explains, this is one aspect that renders the interventionist account of causation superior to those that link causal processes to the exchange of a conserved quantity—which could, for example, be momentum or energy. The charge is that such accounts fail both to correctly distinguish causal and noncausal relationships and to explain the practical point of the ability to distinguish them.

5. *Manipulation* or *intervention* in this sense does not require reference to human action.

6. A generalization G relating X-changes to Y-changes is invariant if "G would continue to hold under some intervention that changes the value of X in such a way that, according to G, the value of Y would change—'continue to hold' in the sense that G correctly describes how the value of Y would change under this intervention" (Woodward 2003, 15).

7. It is assumed that during the intervention, all off-path variables are kept fixed. Also, the possibility that the number of Nobel laureates causally affects the national chocolate consumption has to be excluded.

8. Lewis (1986) discusses this issue in terms of *sensitivity*, whereas Woodward (2010) uses the term *stability*. The basic idea is simple: stability is linked to the counterfactual dependence one continues to obtain under a large number of changes in background circumstances (see Woodward 2010, 291–292).

9. In real-life cases, the mapping from X to Y will only rarely be bijective, but the more we approach it the more causal specificity we get.

10. For instance, the release of neurotransmitters is explained by its previous causes, such as the opening of Ca_2+-channels (Craver 2007).

11. See Illari and Williamson 2012 for different characterizations.

12. To compare, the lightbulb in the fridge is a part of the fridge, but it is not a genuine component of the mechanism—or at least not the one that is responsible for the functioning of the fridge.

13. The literature is often neutral on the ontological question whether a mechanism's higher-level properties supervene on its constituents in a reductive or nonreductive manner (Baumgartner and Gebharter 2016).

14. This is a simplified account of supervenience, but nothing more specific is required for our context. For a discussion of different possibilities, see McLaughlin and Bennett 2018.

15. Craver (2007, 153) acknowledges that at least in some cases (e.g., those involving feedback), manipulability relationships can be bidirectional.

16. A cognitivist account would have predicted that solving the task would require a saccade from the model to the resource area and one from the resource to the workspace area.

17. In part due to the difficulties attached to using spatiotemporal criteria, many of the participants in the demarcation debates discussed in chapter 4 resort to nonspatiotemporal criteria (reasons, ownership, gran level).

18. Another objection is that the requirements fail to consider temporal relations between phenomena and components: (i) could be understood as requiring only

spatial parthood between entities but also as requiring temporal parthood between activities (see Kaiser and Krickel 2016).

19. This also means that surgical interventions on the microlevel are possible without macrolevel changes. However, these are uninformative about constitutive relations.

20. Recall that M was limited to relatively stable and specific causal relationships (see section 5.11).

21. Harinen (2014) also points to other possible combinations. For instance, it is possible to manipulate some Φ with respect to Ψ_{out} while simultaneously manipulating the value of Ψ_{in}.

22. Alternatively, Kirchhoff (2015, 2017) rethinks the distinction between causation and constitution and maintains that constitution can be diachronic, just as causation, and that continuous reciprocal causation can qualify as "constitutive causation" (2017, 867). However, this notion is best suited to grasp activities in complex systems. Clark (1997) introduces the notion of continuous reciprocal causation while describing how systems become less aggregative with increasing frequency and complexity of interactions of their parts. In such cases, due to numerous feedback loops, the activity of components both (in part) determines and is determined by other components, making it difficult to specify distinct causal contributions of components (for a discussion, see Wheeler 2005, 260). It rings true that in such cases, offering an adequate explanation by analyzing insulated causal profiles will probably fail. It is not entirely clear, however, whether a system displaying little aggregativity would be stable enough to be amenable to bottom-up and top-down interventions.

23. Eronen (2013) points to a putative conflict between how Craver and Bechtel account for levels and top-down causation. He argues that Craver and Bechtel sometimes end up describing causal interactions between entities on different levels while holding on to the idea that being on the same level is necessary for causal relationships. Moreover, when lacking a solid definition of levels, accounting for top-down causation by invoking the distinction between interlevel and same-level causation is poised to lead to difficulties. Craver seems to be aware of this issue, conceding that "there can be no unique answer to the question of when two items are at the same level" (Craver 2007, 192).

24. Although top-down causation often is thought to be incompatible with the principle of causal closure of the domain of the physical and the principle of explanatory exclusion (Kim 1993), Kistler (2010) argues that partial top-down constraint conceived in a diachronic manner does not violate these principles.

6 Complex Scaffolding

1. At the same time, physically induced experiences of proximity (by putting someone close to the participant) result in perceptions of higher ambient temperature. Moreover, IJzerman and Semin (2010, 871) also found that feeling similar (or different) to another unknown person can influence participants' perception of ambient temperature (see also Szymkow et al. 2013; for similar correlations between warmth and levels of self-reported social affiliative motivation, see Fay and Maner 2012). On mimicry as linked to social exclusion, see Lakin and Chartrand 2012).

2. It should be noted that by examining pupil size, the Kaspar, Krapp, and König (2015) study concludes that washing may lead to a "clean slate" effect that resets the dynamic in participants' morality judgments.

3. Earp et al. (2014) were unable to replicate the results of Zhong and Liljenquist's (2006) study. However, the authors think that theoretical reasons speak in favor of positing such an effect. It is worth mentioning that social psychology and priming research in particular has been affected by replication failures. But such failures in the replication of individual studies do not warrant drawing drastic conlusions (see Bargh 2012). A large number of studies and meta-analyses indicate that priming is a valid tool (Cameron, Brown-Iannuzzi, and Payne 2012; DeCoster and Claypool 2004), while some point out that priming effects should be expected to exhibit a wide range of variations (Cesario 2014).

4. Inconsistent with the research presented in this section, a recent study conducted with samples of Brazilian students did not find that the experience of weight had significant effects on importance judgments (Rabelo et al. 2015). However, the authors emphasize that the results might show only that the effects are not cross-culturally generalizable.

5. The chapter is restricted to objections that assist calibration efforts and remains thus neutral to the more far-reaching social, ethical, and metaphilosophical claims that Lakoff and Johnson (1999) put forward. For more comprehensive discussions, see Pinker 2007; Gibbs 2011; and Varga 2018a, 2018b.

6. This solves or at least significantly reduces the impact of additional concerns about conceptual metaphors being ambiguous, such that they allow for post hoc explanations for an overly large number of empirical findings (Lakens 2014; Ritchie 2013).

7 Intersomatic Scaffolding

1. Although the literature on emotion regulation is burgeoning, to what extent the processes that achieve the regulation are distinct from the processes that are constitutive of emotions is under debate (Gross and Barrett 2011).

2. I say *qualitative* to exclude general motor behavior changes such as increased muscular activity through exercise, which also may help elevate mood and alleviate emotional suffering.

3. Although they are closely related skills that share neurological foundations; see Chartrand and Lakin 2013.

4. It should be noted that it is not entirely settled whether behavioral matching or imitation provide the only suitable explanation.

5. This may lay the groundwork for the ability to understand other human beings as "like me" (Meltzoff and Moore 1995; Meltzoff and Brooks 2001; Goldman 2000; Tomasello 1999), together with findings that infants inherit a neural mirror system that underlies imitative behavior and forms of social understanding (Umilta et al. 2001; Iacoboni and Dapretto 2006). A potential issue is that the mirror neuron system may develop from sensorimotor learning and thus may not be innate (Ray and Heyes 2011).

6. One might perhaps be reluctant to accept this explanation because it appears to do away with the contrast between causal and constitutive metaphysical relationships. Some have argued that such a line of reasoning is informed by a conventional notion of material constitution, which may not be appropriate in this context. Instead of using this notion of constitution, it has been suggested that we operate with a diachronic notion of constitution (Kirchhoff 2015, 2014). Unlike standard accounts of material constitution, temporality plays a crucial role in understanding dynamically unfolding extended cognitive processes. If constitution also can be comprehended as a diachronic relation analogous to causation, then differences between diachronically developing relations of constitution and "continuous reciprocal causation" largely lose their significance.

8 Scaffolded Social Cognition

1. As Uta Frith (2012, 2079) notes in retrospect, "In my enthusiasm for the new theory, I frequently overlooked that mentalizing problems can only account for a fraction of social impairments. Social cognition has many more tricks up its sleeve than mentalizing."

2. The same capacity is, however, sometimes referred to as *mentalizing* or *theory of mind*.

3. See Gallagher and Varga 2015b for a discussion of social cognition and mental disorders, and see Gallagher and Varga 2015a for conceptual questions in ASD.

4. For an early iteration of this idea, see Sellars 1956.

5. A thorough discussion of this issue lies beyond the scope of this chapter, but it is worth noting that some have questioned whether *theorizing* is the right term to

describe such processes (see Gallagher 2007). Theorizing indeed seems to suggest a person-level activity.

6. Although the relevant mechanism seems to lack the kind of universality and encapsulation from exogenous social factors that usually characterizes modules, some argue that the very fact of being a spectrum disorder renders the nonmodular explanation of mindreading impairments less plausible than a modular explanation (Adams 2013). Ultimately, support for a domain-specific explanation would require studies indicating a mindreading deficit without deficits in, for instance, communication, imagination, and motor behavior.

7. However, proponents sometimes maintain that interaction theory is best seen as an alternative to mindreading. This has led to several objections (see Michael 2011). This chapter accepts that embodied interaction can support social understanding, while maintaining that mindreading is often required for social understanding. The investigation of embodied, low-level social cognition is thus best thought of as complementing traditional research on higher-level processes that track mental states.

8. Kanner (1943) had already noted motor abnormalities in gait and anticipatory postures.

9. Teitelbaum et al. (1998) provided a still-frame analysis of videos taken in infancy of seventeen children who were later, at the age of three, diagnosed with ASD. The authors reported that all children displayed disturbances of movement revealed in developmental milestones such as rotating, crawling, sitting, and walking.

10. Studies using galvanic skin conductance have revealed decreased autonomic arousal when adults with ASD view faces expressing emotion (Hubert et al. 2009).

11. It should be added that the studies cannot rule out that children with ASD pay less attention to their parents on the whole.

References

Ackerman, J. M., C. C. Nocera, and J. A. Bargh. 2010. "Incidental Haptic Sensations Influence Social Judgments and Decisions." *Science* 328 (5986): 1712–1715.

Adams, F. 2000. "Asymmetrical Dependence" In *A Field Guide to the Philosophy of Mind*, edited by M. Nani and M. Marrafa. http://host.uniroma3.it/progetti/kant/field/.

Adams, F., and K. Aizawa. 2008. *The Bounds of Cognition*. Oxford: Blackwell.

Adams, F., and K. Aizawa. 2010. *The Bounds of Cognition*. Oxford: Wiley-Blackwell.

Adams, F., and R. Garrison. 2013. "The Mark of the Cognitive." *Minds and Machines* 23 (3): 339–352.

Adams, M. 2013. "Explaining the Theory of Mind Deficit in Autism Spectrum Disorder." *Philosophical Studies* 163 (1): 233–249.

Adams, R. A., Q. J. Huys, and J. P. Roiser. 2015. "Computational Psychiatry: Towards a Mathematically Informed Understanding of Mental Illness." *Journal of Neurology, Neurosurgery, and Psychiatry* 87: 53–63.

Aldao, A., and S. Nolen-Hoeksema. 2012a. "The Influence of Context on the Implementation of Adaptive Emotion Regulation Strategies." *Behaviour Research and Therapy* 50 (7): 493–501.

Aldao, A., and S. Nolen-Hoeksema. 2012b. "When Are Adaptive Strategies Most Predictive of Psychopathology?" *Journal of Abnormal Psychology* 121 (1): 276.

Aldao, A., S. Nolen-Hoeksema, and S. Schweizer. 2010. "Emotion-Regulation Strategies across Psychopathology: A Meta-analytic Review." *Clinical Psychology Review* 30 (2): 217–237.

Alford, B. A., and A. T. Beck. 1997. "The Relation of Psychotherapy Integration to the Established Systems of Psychotherapy." *Journal of Psychotherapy Integration* 7 (4): 275.

Anderson, M. L. 2003. "Embodied Cognition: A Field Guide." *Artificial Intelligence* 149 (1): 91–130.

Anderson, M. L. 2010. "Neural Reuse: A Fundamental Organizational Principle of the Brain." *Behavioral and Brain Sciences* 33 (4): 245–266.

Anderson, M. L. 2014. *After Phrenology: Neural Reuse and the Interactive Brain*. Cambridge, MA: MIT Press.

Anderson, M. L., M. J. Richardson, and A. Chemero. 2012. "Eroding the Boundaries of Cognition: Implications of Embodiment." *Topics in Cognitive Science* 4 (4): 717–730.

Andres, M., E. Olivier, and A. Badets. 2008. "Actions, Words, and Numbers: A Motor Contribution to Semantic Processing? Current Direction in Psychology." *Science* 17:313–317.

Apperly, I. A. 2011. *Mindreaders: The Cognitive Basis of "Theory of Mind."* New York: Psychology Press.

Apperly, I. A., and S. A. Butterfill. 2009. "Do Humans Have Two Systems to Track Beliefs and Belief-Like States?" *Psychological Review* 116 (4): 953.

Baars, B. J. 1986. *The Cognitive Revolution in Psychology*. New York: Guilford Press.

Baillargeon, R., R. M. Scott, and Z. He. 2010. "False-Belief Understanding in Infants." *Trends in Cognitive Sciences* 14 (3): 110–118.

Ballard, D. H., M. M. Hayhoe, and J. B. Pelz. 1995. "Memory Representations in Natural Tasks." *Journal of Cognitive Neuroscience* 7 (1): 66–80.

Ballard, D. H., M. M. Hayhoe, P. K. Pook, and R. P. N. Rao. 1997. "Deictic Codes for the Embodiment of Cognition." *Behavioral and Brain Sciences* 20:723–767.

Bargh, J. A. 2012. "Priming Effects Replicate Just Fine, Thanks." *Psychology Today* (blog). May 11. https://www.psychologytoday.com/us/blog/the-natural-unconscious/201205/priming-effects-replicate-just-fine-thanks.

Bargh, J. A., and I. Shalev. 2012. "The Substitutability of Physical and Social Warmth in Daily Life." *Emotion* 12 (1): 154–162.

Baron-Cohen, S. 1997. *Mindblindness: An Essay on Autism and Theory of Mind*. Cambridge, MA: MIT Press.

Baron-Cohen, S. 2008. "Theories of the Autistic Mind." *Psychologist* 21:112–116.

Baron-Cohen S., A. Leslie, and U. Frith. 1985. "Does the Autistic Child Have a Theory of Mind?" *Cognition* 21:37–46.

Barrett, L. 2011. *Beyond the Brain: How Body and Environment Shape Animal and Human Minds*. Princeton, NJ: Princeton University Press.

Barsalou, L. W. 1999. "Perceptions of Perceptual Symbols." *Behavioral and Brain Sciences* 22 (4): 637–660.

Barsalou, L. W. 2008. "Grounded Cognition." *Annual Review of Psychology* 59 (1): 617–645.

Barsalou, L. W. 2010. "Grounded Cognition: Past, Present, and Future: Topics in Cognitive Science." *Topics in Cognitive Science* 2 (4): 716–724.

Baumeister, J. C., G. Papa, and F. Foroni. 2016. "Deeper than Skin Deep: The Effect of Botulinum Toxin-A on Emotion Processing." *Toxicon* 118:86–90.

Baumgartner, M., and L. Casini. 2017. "An Abductive Theory of Constitution." *Philosophy of Science* 84 (2): 214–233.

Baumgartner, M., and A. Gebharter. 2016. "Constitutive Relevance, Mutual Manipulability, and Fat-Handedness." *British Journal for the Philosophy of Science* 67 (3): 731–756.

Baumgartner, M., and W. Wilutzky. 2017. "Is It Possible to Experimentally Determine the Extension of Cognition?" *Philosophical Psychology* 30 (8): 1104–1125.

Beall, P. M., E. J. Moody, D. N. McIntosh, S. L. Hepburn, and C. L. Reed. 2008. "Rapid Facial Reactions to Emotional Facial Expressions in Typically Developing Children and Children with Autism Spectrum Disorder." *Journal of Experimental Child Psychology* 101 (3): 206–223.

Bechara, A., H. Damasio, D. Tranel, and A. R. Damasio. 1997. "Deciding Advantageously before Knowing the Advantageous Strategy." *Science* 275 (5304): 1293–1295.

Bechara, A., H. Damasio, D. Tranel, and A. R. Damasio. 2005. "The Iowa Gambling Task and the Somatic Marker Hypothesis: Some Questions and Answers." *Trends in Cognitive Sciences* 9 (4): 159–162.

Bechtel, W. 2008. *Mental Mechanisms: Philosophical Perspectives on Cognitive Neuroscience*. London: Routledge.

Bechtel, W. 2009. "Constructing a Philosophy of Science of Cognitive Science." *Topics in Cognitive Science* 1 (3): 548–569.

Bechtel, W. 2010. "How Can Philosophy Be a True Cognitive Science Discipline?" *Topics in Cognitive Science* 2 (3): 357–366.

Bechtel, W. 2017. "Explicating Top-Down Causation Using Networks and Dynamics." *Philosophy of Science* 84 (2): 253–274.

Bechtel, W. and A. Abrahamsen. 2002. *Connectionism and the Mind*. Malden, MA: Blackwell.

Bechtel, W., and A. Abrahamsen. 2005. "Explanation: A Mechanist Alternative." *Studies in History and Philosophy of Science Part C: Studies in History and Philosophy of Biological and Biomedical Sciences* 36 (2): 421–441.

Bechtel, W., and M. Herschbach. 2010. Philosophy of the Cognitive Sciences. In *Philosophy of the Sciences*, edited by F. Allhoff, 239–261. Oxford: Blackwell.

Bechtel, W., and R. C. Richardson. 2010. *Discovering Complexity: Decomposition and Localization as Strategies in Scientific Research*. Cambridge, MA: MIT Press.

Beck, A. T. 1976. *Cognitive Therapy and the Emotional Disorders*. New York: International Universities Press.

Beck, A. T., and B. A. Alford. 2009. *Depression: Causes and Treatment*. Philadelphia: University of Pennsylvania Press.

Beck, A. T., A. Freeman, J. Pretzer, D. D. Davis, B. Fleming, R. Ottavani, J. Beck, K. M. Simon, et al. 1990. *Cognitive Therapy of Personality Disorders*. New York: Guilford Press.

Beck, A. T., and M. E. Weishaar. 2008. "Cognitive Therapy." In *Current Psychotherapies*, edited by R. J. Corsini and D. Wedding, 263–292. Belmont, CA: Thomson.

Beck, J. S. 1995. *Cognitive Therapy: Basics and Beyond*. New York: Guilford Press.

Beeghly, M., and E. Tronick. 2011. "Early Resilience in the Context of Parent–Infant Relationships: A Social Developmental Perspective." *Current Problems in Pediatric and Adolescent Health Care* 41 (7): 197–201.

Beer, R. D. 1995. "A Dynamical Systems Perspective on Agent-Environment Interaction." *Artificial Intelligence* 72 (1–2): 173–215.

Beer, R. D. 2003. "The Dynamics of Active Categorical Perception in an Evolved Model Agent." *Adaptive Behavior* 11 (4): 209–243.

Beer, R. D. 2014. "Dynamical Systems and Embedded Cognition." In *The Cambridge Handbook of Artificial Intelligence*, no. 812: 856–873.

Bennabi, D., P. Vandel, C. Papaxanthis, T. Pozzo, and E. Haffen. 2013. "Psychomotor Retardation in Depression: A Systematic Review of Diagnostic, Pathophysiologic, and Therapeutic Implications." *BioMed Research International* 2013:1–18. https://doi.org/10.1155/2013/158746.

Berkeley, I. S. N. 2008. "CajunBot: A Case Study of an Embodied System." In *The Elsevier Handbook of Embodied Cognitive Science*, edited by P. C. Garzon and A. Gomila, 81–98. Oxford: Elsevier Ltd.

Berking, M., and B. Whitley. 2014. "Emotion Regulation: Definition and Relevance for Mental Health." In *Affect Regulation Training*, edited by M. Berking and B. Whitley, 5–17. New York: Springer.

Bermúdez, J. L. 2014. *Cognitive Science: An Introduction to the Science of the Mind*. Cambridge: Cambridge University Press.

Bernieri, F. J., J. M. Davis, R. Rosenthal, and C. R. Knee. 1994. "Interactional Synchrony and Rapport: Measuring Synchrony in Displays Devoid of Sound and Facial Affect." *Personality and Social Psychology Bulletin* 20 (3): 303–311.

Bernieri, F. J., and R. Rosenthal 1991. "Interpersonal Coordination: Behavior Matching and Interactional Synchrony." In *Studies in Emotion and Social Interaction: Fundamentals of Nonverbal Behavior*, edited by R. S. Feldman and B. Rimé, 401–432. New York: Cambridge University Press.

Bhat, A. N., R. J. Landa, and J. C. Galloway. 2011. "Current Perspectives on Motor Functioning in Infants, Children, and Adults with Autism Spectrum Disorders." *Physical Therapy* 91 (7): 1116–1129.

Binder, J. R., and R. H. Desai. 2011. "The Neurobiology of Semantic Memory." *Trends in Cognitive Sciences* 15 (11): 527–536.

Blatt, S. J. 2004. *Experiences of Depression: Theoretical, Clinical, and Research Perspectives*. Washington, DC: American Psychological Association.

Block, N. 1991. "Troubles with Functionalism." In *The Nature of Mind*, edited by D. Rosenthal, 211–228. Oxford, UK: Oxford University Press.

Bly, B. M., and D. E. Rumelhart. 1999. *Cognitive Science*. San Diego, CA: Academic Press.

Boden, M. A. 2006. *Mind as Machine: A History of Cognitive Science*. New York: Oxford University Press.

Borg, J. S., D. Lieberman, and K. A. Kiehl. 2008. "Infection, Incest, and Iniquity: Investigating the Neural Correlates of Disgust and Morality." *Journal of Cognitive Neuroscience* 20:1529–1546.

Bortolotti, L., and M. Broome. 2009. "The Future of Scientific Psychiatry." In *Psychiatry as Cognitive Neuroscience: Philosophical Perpsctives*, edited by M. Broome and L. Bortolotti, 365–375. Oxford: Oxford University Press.

Boyle, G. J., L. Stankov, and R. B. Cattell. 1995. "Measurement and Statistical Models in the Study of Personality and Intelligence." In *International Handbook of Personality and Intelligence*, edited by D. H. Saklofski and M. Zeidner, 417–446. New York: Springer.

Boyd, R. 1991. "Realism, Anti-foundationalism and the Enthusiasm for Natural Kinds." *Philosophical Studies* 61 (1–2): 127–148.

Branigan, H. P., M. J. Pickering, and A. A. Cleland. 2000. "Syntactic Co-ordination in Dialogue." *Cognition* 75 (2): B13–B25.

Brincker, M., and E. B. Torres. 2013. "Noise from the Periphery in Autism." *Frontiers in Integrative Neuroscience* 7:34. https://doi.org/10.3389/fnint.2013.00034.

Brook, A. 2009. "Introduction: Philosophy in and Philosophy of Cognitive Science." *Topics in Cognitive Science* 1 (2): 216–230.

Brooks, R. A. 1991. "Intelligence without Representation." *Artificial Intelligence* 47 (1–3): 139–159.

Brooks, R. A. 1999. *Cambrian Intelligence: The Early History of the New AI*. Cambridge, MA: MIT Press.

Brüne, M., and U. Brüne-Cohrs. 2006. "Theory of Mind: Evolution, Ontogeny, Brain Mechanisms and Psychopathology." *Neuroscience and Biobehavioral Reviews* 30 (4): 437–455.

Brüne, M., C. Sonntag, M. Abdel-Hamid, C. Lehmkämper, G. Juckel, and A. Troisi. 2008. "Nonverbal Behavior during Standardized Interviews in Patients with Schizophrenia Spectrum Disorders." *The Journal of Nervous and Mental Disease* 196 (4): 282–288.

Bruner, J. 1975. "From Communication to Language: A Psychological Perspective." *Cognition* 3:255–287.

Buccino, G., F. Binkofski, G. R. Fink, L. Fadiga, L. Fogassi, V. Gallese, R. J. Seitz, K. Zilles, G. Rizzolatti, and H.-J. Freund. 2001. "Action Observation Activates Premotor and Parietal Areas in a Somatotopic Manner: An fMRI Study." *European Journal of Neuroscience* 13 (2): 400–404.

Buckner, C. 2015. "A Property Cluster Theory of Cognition." *Philosophical Psychology* 28 (3): 307–336.

Burge, T. 1979. "Individualism and the Mental." *Midwest Studies in Philosophy* 4 (1): 73–121.

Butner, J., L. M. Diamond, and A. M. Hicks. 2007. "Attachment Style and Two Forms of Affect Coregulation between Romantic Partners." *Personal Relationships* 14 (3): 431–455.

Buyukdura, J. S., S. M. McClintock, and P. E. Croarkin. 2011. "Psychomotor Retardation in Depression: Biological Underpinnings, Measurement, and Treatment." *Progress in Neuro-Psychopharmacology and Biological Psychiatry* 35 (2): 395–409.

Caligiuri, M. P., and J. Ellwanger. 2000. "Motor and Cognitive Aspects of Motor Retardation in Depression." *Journal of Affective Disorders* 57:83–93.

Call, J., and M. Tomasello. 1998. "Distinguishing Intentional from Accidental Actions in Orangutans (*Pongo pygmaeus*), Chimpanzees (*Pan troglodytes*), and Human Children (*Homo sapiens*)." *Journal of Comparative Psychology* 112 (2): 196–206.

Calugi, S., G. B. Cassano, A. Litta, P. Rucci, A. Benvenuti, M. Miniati, A. Fagiolini, et al. 2011. "Does Psychomotor Retardation Define a Clinically Relevant Phenotype of Unipolar Depression?" *Journal of Affective Disorders* 129 (1): 296–300.

Cannizzaro, M., B. Harel, N. Reilly, P. Chappell, and P. J. Snyder. 2004. "Voice Acoustical Measurement of the Severity of Major Depression." *Brain and Cognition* 56 (1): 30–35.

Cannon, P. R., S. Schnall, and M. White. 2010. "Transgressions and Expressions: Affective Facial Muscle Activity Predicts Moral Judgments." *Social Psychological and Personality Science* 2:325–331.

Cardona, J. F., L. Kargieman, V. Sinay, O. Gershanik, C. Gelormini, L. Amoruso, and A. Ibáñez. 2014. "How Embodied Is Action Language? Neurological Evidence from Motor Diseases." *Cognition* 131:311–322.

Carlson, R. A., M. N. Avraamides, M. Cary, and S. Strasberg. 2007. "What Do the Hands Externalize in Simple Arithmetic?" *Journal of Experimental Psychology: Learning, Memory, and Cognition* 33 (4): 747–756.

Carruthers, P. 1996. "Simulation and Self-Knowledge: A Defense of Theory-Theory." In *Theories of Theories of Mind*, edited by P. Carruthers and P. K. Smith, 22–68. Cambridge: Cambridge University Press.

Carruthers, P. 2015. "Perceiving Mental States." *Consciousness and Cognition* 36:498–507.

Casasanto, D. 2013. "Experiential origins of mental metaphors: Language, culture, and the body. In *The Power of Metaphor: Examining its Influence on Social Life*, edited by M. Landau, M. D. Robinson, and B. Meier, 246–268. Washington, DC: American Psychological Association Books.

Casasanto, D. 2014. "Development of Metaphorical Thinking: The Role of Language." In *Language and the Creative Mind*, edited by M. Borkent, B. Dancygier, and J. Hinnell, 3–18. Stanford, CA: CSLI Publications.

Casasanto, D., and L. Boroditsky. 2008. "Time in the Mind: Using Space to Think about Time." *Cognition* 106:579–593.

Caspar, F., T. Rothenfluh, and Z. Segal. 1992. "The Appeal of Connectionism for Clinical Psychology." *Clinical Psychology Review* 12 (7): 719–762.

Caspers, S., K. Zilles, A. R. Laird, and S. B. Eickhoff. 2010. "ALE Meta-analysis of Action Observation and Imitation in the Human Brain." *Neuroimage* 50 (3): 1148–1167.

Cattaneo, L., G. Barchiesi, D. Tabarelli, C. Arfeller, M. Sato, and A. M. Glenberg. 2011. "One's Motor Performance Predictably Modulates the Understanding of

Others' Actions through Adaptation of Premotor Visuo-motor Neurons." *Social Cognitive and Affective Neuroscience* 6:301–310.

Cattaneo, L., F. Caruana, A. Jezzini, and G. Rizzolatti. 2009. "Representation of Goal and Movements without Overt Motor Behavior in the Human Motor Cortex: A Transcranial Magnetic Stimulation Study." *Journal of Neuroscience* 29 (36): 11134–11138.

Cattaneo, L., M. Fabbri-Destro, S. Boria, C. Pieraccini, A. Monti, G. Cossu, and G. Rizzolatti. 2007. "Impairment of Actions Chains in Autism and Its Possible Role in Intention Understanding." *Proceedings of the National Academy of Sciences* 104 (45): 17825–17830.

Chakravartty, A. 2011. "Scientific Realism and Ontological Relativity." *Monist* 94 (2): 157–180.

Chalmers, D. 1993. "Connectionism and Compositionality: Why Fodor and Pylyshyn Were Wrong." *Philosophical Psychology* 6:305–319.

Chalmers, D. 2008. Foreword to *Supersizing the Mind: Embodiment, Action, and Cognitive Extension,* by A. Clark, ix–xvi. New York: Oxford University Press.

Chang, H. 2012. *Is Water H2O? Evidence, Realism and Pluralism.* Dordrecht: Springer.

Chapman, S. T. A., D. A. Kim, J. M. Susskind, and A. K. Anderson. 2009. "In Bad Taste: Evidence for the Oral Origins of Moral Disgust." *Science* 323:1222–1226.

Chartrand, T. L., and J. L. Lakin. 2013. "The Antecedents and Consequences of Human Behavioral Mimicry." *Annual Review of Psychology* 64:285–308.

Chemero, A. 2009. *Radical Embodied Cognitive Science.* Cambridge, MA: MIT Press.

Chemero, A. 2013. "Radical Embodied Cognitive Science." *Review of General Psychology* 17 (2): 145–150.

Chen J., L. Yang, W. Ma, X. Wu, Y. Zhang, and D. Wei. 2013. "Ego-Rotation and Object-Rotation in Major Depressive Disorder." *Psychiatry Research* 209 (1): 32–39.

Chu, M., and S. Kita. 2011. "The Nature of Gestures' Beneficial Role in Spatial Problem Solving." *Journal of Experimental Psychology: General* 140 (1): 102–116.

Churchland, P. M. 1995. *The Engine of Reason, the Seat of the Soul.* Cambridge, MA: MIT Press.

Churchland, P. M., and P. S. Churchland. 1990. "Could a Machine Think?" *Scientific American* 262 (1): 32–37.

Churchland, P. S., and T. Sejnowski. 1992. *The Computational Brain.* Cambridge, MA: MIT Press.

Clark, A. 1989. *Microcognition: Philosophy, Cognitive Science, and Parallel Distributed Processing*. Cambridge, MA: MIT Press.

Clark, A. 1997. *Being There: Putting Brain, Body, and World Together Again*. Cambridge, MA: MIT Press.

Clark, A. 1999. "An Embodied Cognitive Science?" *Trends in Cognitive Sciences* 3 (9): 345–351.

Clark, A. 2007. "Re-inventing Ourselves: The Plasticity of Embodiment, Sensing, and Mind." *Journal of Medicine and Philosophy* 32 (3): 263–282.

Clark, A. 2008a. "Pressing the Flesh: A Tension in the Study of the Embodied, Embedded Mind?" *Philosophy and Phenomenological Research* 76 (1): 37–59.

Clark, A. 2008b. *Supersizing the Mind: Embodiment, Action, and Cognitive Extension*. New York: Oxford University Press.

Clark, A. 2009. "Spreading the Joy: Why the Machinery of Consciousness Is (Probably) Still in the Head." *Mind* 118 (472): 963–993.

Clark, A. 2010a. "Memento's Revenge: The Extended Mind, Extended." In *The Extended Mind*, edited by R. Menary, 43–66. Cambridge, MA: MIT Press.

Clark A. 2010b. "Coupling, Constitution, and the Cognitive Kind: A Reply to Adams and Aizawa." In *The Extended Mind*, edited by R. Menary, 81–99. Cambridge, MA: MIT Press.

Clark, A. 2011. "Finding the Mind." *Philosophical Studies* 152 (3): 447–461.

Clark, A. 2014. *Mindware: An Introduction to the Philosophy of Cognitive Science*. Oxford: Oxford University Press.

Clark, A., and D. Chalmers. 1998. "The Extended Mind." *Analysis* 58:10–23.

Clark, A., and J. Toribio. 1994. "Doing without Representing?" *Synthese* 101 (3): 401–431.

Coan, J. A. 2010. "Adult Attachment and the Brain." *Journal of Social and Personal Relationships* 27 (2): 210–217.

Coan, J. A., and E. L. Maresh. 2014. "Social Baseline Theory and the Social Regulation of Emotion." In *Handbook of Emotion Regulation*, edited by J. J. Gross, 221–236. New York: Guilford Press.

Cochin, S., C. Barthelemy, and S. Roux. 1999. "Observation and Execution of Movement: Similarities Demonstrated by Quantified Electroencephalography." *European Journal of Neuroscience* 11:1839–1842.

Colby, K. M. 1981. "Modeling a Paranoid Mind." *Behavioral and Brain Sciences* 4 (4): 515–534.

Colby, K. M., S. Weber, and F. D. Hilf. 1971. "Artificial Paranoia." *Artificial Intelligence* 2 (1): 1–25.

Cole, P. M., S. E. Hall, and A. M. Radzioch. 2009. "Emotional Dysregulation and the Development of Serious Misconduct." In *Biopsychosocial Regulatory Processes in the Development of Childhood Behavioral Problems*, edited by S. L. Olson and A. J. Sameroff, 186–211. New York: Cambridge University Press.

Colombetti, G., and J. Krueger. 2015. "Scaffoldings of the Affective Mind." *Philosophical Psychology* 28 (8): 1157–1176.

Colombi, C., G. Vivanti, and S. Rogers. 2011. "The Neuropsychology of Imitation Deficit in Autism." In *The Neuropsychology of Autism*, edited by D. Fein, 243–266. Oxford: Oxford University Press.

Cook, J., S. Blakemore, and C. Press. 2013. "Atypical Basic Movement Kinematics in Autism Spectrum Conditions." *Brain* 136:2816–2824.

Cosmides, L., and J. Tooby. 1992. "Cognitive Adaptations for Social Exchange." In *The Adapted Mind: Evolutionary Psychology and the Generation of Culture*, edited by J. H. Barkow, L. Cosmides, and J. Tooby, 163–228. New York: Oxford University Press.

Cratsley, K., and R. Samuels. 2013. "Cognitive Science and Explanations of Psychopathology." In *The Oxford Handbook of Philosophy and Psychiatry*, edited by K.W. M. Fulford, M. Davies, R. G. T. Gipps, G. Graham, J. Z. Sadler, G. Stanghellini, and T. Thornton, 413–433. Oxford: Oxford University Press.

Craver, C. F. 2007. *Explaining the Brain: Mechanisms and the Mosaic Unity of Neuroscience*. Oxford: Oxford University Press.

Craver, C. F. 2009. "Mechanisms and Natural Kinds." *Philosophical Psychology* 22 (5): 575–594.

Craver, C. F., and W. Bechtel. 2007. "Top-Down Causation without Top-Down Causes." *Biology & Philosophy* 22 (4): 547–563.

Craver, C. F., and L. Darden. 2013. *In Search of Mechanisms: Discoveries across the Life Sciences*. Chicago: University of Chicago Press.

Craver, C., and J. Tabery. 2017. "Mechanisms in Science." In *The Stanford Encyclopedia of Philosophy*, edited by E. N. Zalta. https://plato.stanford.edu/archives/spr2017/entries/sciencemechanisms.

Cummins, Robert. 1998. "Reflection on Reflective Equilibrium." In *Rethinking Intuition*, edited by M. DePaul and W. Ramsey, 113–128. Lanham, MD: Rowman & Littlefield.

Dale, R. 2008. "The Possibility of a Pluralist Cognitive Science." *Journal of Experimental and Theoretical Artificial Intelligence* 20 (3): 155–179.

Dale, R., E. Dietrich, and A. Chemero. 2009. "Explanatory Pluralism in Cognitive Science." *Cognitive Science* 33 (5): 739–742.

Dale, R., and M. J. Spivey. 2006. "Unraveling the Dyad: Using Recurrence Analysis to Explore Patterns of Syntactic Coordination between Children and Caregivers in Conversation." *Language Learning* 56 (3): 391–430.

Dalgleish, T., and F. N. Watts. 1990. "Biases of Attention and Memory in Disorders of Anxiety and Depression." *Clinical Psychology Review* 10 (5): 589–604.

Damasio, A. R. 1994. *Descartes' Error: Emotion, Rationality and the Human Brain.* New York Grosset/Putnam.

Damasio, A. R. 1999. *The Feeling of What Happens: Body, Emotion and the Making of Consciousness.* London: Heinemann.

Damasio, A. R. 2003. *Looking for Spinoza: Joy, Sorrow, and the Feeling Brain.* New York: Houghton Mifflin Harcourt.

Damasio, A. R., and R. G. Maurer. 1978. "A Neurological Model for Childhood Autism." *Archives of Neurology* 35:777–786.

Dapretto, M., M. S. Davies, J. H. Pfeifer, A. A. Scott, M. Sigman, S. Y. Bookheimer, and M. Iacoboni. 2006. "Understanding Emotions in Others: Mirror Neuron Dysfunction in Children with Autism Spectrum Disorders." *Nature Neuroscience* 9 (1): 28–30.

Darwin, C. 1862. *On the Various Contrivances by which British and Foreign Orchids Are Fertilized by Insects, and on the Good Effects of Intercrosssing.* London: John Murray.

Davies, M. 2005. "Cognitive Science." In *The Oxford Handbook of Contemporary Philosophy*, edited by F. Jackson and M. Smith, 358–394. New York: Oxford University Press.

Dawkins, R. 1976. *The Selfish Gene.* New York: Oxford University Press.

Day, M. V., and D. R. Bobocel. 2013. "The Weight of a Guilty Conscience: Subjective Body Weight as an Embodiment of Guilt." *PLOS ONE* 8:1–7

Dehaene, S. 2009. *Reading in the Brain: The New Science of How We Read.* New York: Penguin.

Delaherche, E., M. Chetouani, A. Mahdhaoui, C. Saint-Georges, S. Viaux, and D. Cohen. 2012. "Interpersonal Synchrony: A Survey of Evaluation Methods across Disciplines." *IEEE Transactions on Affective Computing* 3 (3): 349–365.

Denke, C., M. Rotte, H.-J. Heinze, and M. Schaefer. 2014. "Lying and the Subsequent Desire for Toothpaste: Activity in the Somatosensory Cortex Predicts Embodiment of the Moral-Purity Metaphor." *Cerebral Cortex* 26 (2): 477–484.

Dennett, D. C. 1987. *The Intentional Stance*. Cambridge, MA: MIT Press.

Dennett, D. C. 1996. *Kinds of Minds: Toward an Understanding of Consciousness*. New York: Basic Books.

Dennett, D. C. 2009. "The Part of Cognitive Science That Is Philosophy." *Topics in Cognitive Science* 1 (2): 231–236.

Dennett, D. C. 1995. *Darwin's Dangerous Idea: Evolution and the Meanings of Life*. New York: Simon and Schuster

DeWall, C. N., G. MacDonald, G. D. Webster, C. L. Masten, R. F. Baumeister, C. Powell, D. Combs, et al. 2010. "Acetaminophen Reduces Social Pain: Behavioral and Neural Evidence." *Psychological Science* 21:931–937.

DeWall, C. N., and R. F. Baumeister. 2006. "Alone but Feeling No Pain: Effects of Social Exclusion on Physical Pain Tolerance and Pain Threshold, Affective Forecasting, and Interpersonal Empathy." *Journal of Personality and Social Psychology* 91 (1): 1–15.

DeWitt, Michael. 2006. "Intuitions in Linguistics." *British Journal for the Philosophy of Science* 57 (3): 481–513.

Di Paolo, E., M. Rohde, and H. De Jaegher. 2010. "Horizons for the Enactive Mind: Values, Social Interaction, and Play." In *Enaction: Towards a New Paradigm for Cognitive Science*, edited by J. Stewart, O. Gapenne, and E. Di Paolo, 33–38. Cambridge, MA: MIT Press.

Dimic, S., C. Wildgrube, R. McCabe, I. Hassan, T. R. Barnes, and S. Priebe. 2010. "Non-verbal Behaviour of Patients with Schizophrenia in Medical Consultations: A Comparison with Depressed Patients and Association with Symptom Levels." *Psychopathology* 43 (4): 216–222.

Dowd, A. M., N. J. Rinehart, and J. McGinley. 2010. "Motor Function in Children with Autism: Why Is This Relevant to Psychologists?" *Clinical Psychologist* 14 (3): 90–96.

Drayson, Z. 2009. "Embodied Cognitive Science and Its Implications for Psychopathology." *Philosophy, Psychiatry, and Psychology* 16 (4): 329–340.

Drayson, Z. 2010. "Extended Cognition and the Metaphysics of Mind." *Cognitive Systems Research* 11 (4): 367–377.

Drayson, Z., and A. Clark. 2013. "Augmentation, Agency, and the Spreading of the Mental State." Forthcoming in *Consciousness and Cognition*.

DSM-5. 2013. *Diagnostic and Statistical Manual of Mental Disorders*. 5th ed. Arlington, VA: American Psychiatric Association.

Duncan, S., and L. F. Barrett. 2007. "Affect Is a Form of Cognition: A Neurobiological Analysis." *Cognition and Emotion* 21 (6): 1184–1211.

Dupré, J. 1993. *The Disorder of Things*. Cambridge, MA: Harvard University Press.

Earp, B. D., J. A. Everett, E. N. Madva, and J. K. Hamlin. 2014. "Out, Damned Spot: Can the 'Macbeth Effect' Be Replicated?" *Basic and Applied Social Psychology* 36 (1): 91–98.

Ehring, T., S. Fischer, J. Schnülle, A. Bösterling, and B. Tuschen-Caffier. 2008. "Characteristics of Emotion Regulation in Recovered Depressed versus Never Depressed Individuals." *Personality and Individual Differences* 44 (7): 1574–1584.

Eigsti, I.-M. 2013. "A Review of Embodiment in Autism Spectrum Disorders." *Frontiers in Psychology* 4:224.

Eisenberger, N. I. 2012. "Broken Hearts and Broken Bones: A Neural Perspective on the Similarities between Social and Physical Pain." *Current Directions in Psychological Science* 21:42–47.

Eisenberger, N. I., M. D. Lieberman, and K. D. Williams. 2003. "Does Rejection Hurt? An fMRI Study of Social Exclusion." *Science* 302 (5643): 290–292.

Elfenbein, H. A., M. G. Beaupré, M. Lévesque, and U. Hess. 2007. "Toward a Dialect Theory: Cultural Differences in the Expression and Recognition of Posed Facial Expressions." *Emotion* 7:131–146.

Elman, J. L. 1991. "Distributed Representations, Simple Recurrent Networks, and Grammatical Structure." *Machine Learning* 7 (2–3): 195–225.

Elman, J. L. 1993. "Learning and Development in Neural Networks: The Importance of Starting Small." *Cognition* 48 (1): 71–99.

Elpidorou, A. 2014. "Reasoning about the Mark of the Cognitive: A Response to Adams and Garrison." *Minds and Machines* 24:201–211.

Elster, Jakob. 2011. "How Outlandish Can Imaginary Cases Be?" *Journal of Applied Philosophy* 28 (3): 241–258.

Enticott, P. G., H. A. Kennedy, N. J. Rinehart, J. L. Bradshaw, B. J. Tonge, Z. J. Daskalakis, and P. B. Fitzgerald. 2013. "Interpersonal Motor Resonance in Autism Spectrum Disorder: Evidence against a Global Mirror System Deficit." *Frontiers in Human Neuroscience* 7:218.

Epp, A. M., and K. S. Dobson. 2010. "The Evidence Base for Cognitive-Behavioral Therapy." In *Handbook of Cognitive-Behavioral Therapies*, edited by K. S. Dobson, 39–73. New York: Guilford Press.

Erlhagen, W., and G. Schöner. 2002. "Dynamic Field Theory of Movement Preparation." *Psychological Review* 109:545–572.

Eronen, M. I. 2013. "No Levels, No Problems: Downward Causation in Neuroscience." *Philosophy of Science* 80 (5): 1042–1052.

Fay, A. J., and J. K. Maner. 2012. "Warmth, Spatial Proximity, and Social Attachment: The Embodied Perception of a Social Metaphor." *Journal of Experimental Social Psychology* 48:1369–1372.

Fehr, B., and J. A. Russell. 1984. "Concept of Emotion Viewed from a Prototype Perspective." *Journal of Experimental Psychology: General* 113 (3): 464–486.

Feldman, R. 2003. "Infant–Mother and Infant–Father Synchrony: The Coregulation of Positive Arousal." *Infant Mental Health Journal* 24 (1): 1–23.

Feldman, R. 2007. "On the Origins of Background Emotions: From Affect Synchrony to Symbolic Expression." *Emotion* 7 (3): 601–611.

Feldman, R. 2015. "Mutual Influences between Child Emotion Regulation and Parent-Child Reciprocity Support Development across the First 10 Years of Life: Implications for Developmental Psychopathology." *Development and Psychopathology* 27 (4, pt. 1): 1007–1023.

Feldman, R. 2017. "The Neurobiology of Human Attachments." *Trends in Cognitive Sciences* 21 (2): 80–99.

Feldman, R., and A. I. Eidelman. 2004. "Parent–Infant Synchrony and the Social-Emotional Development of Triplets." *Developmental Psychology* 40:1133–1147.

Fenech, E. P., F. A. Drews, and J. Z. Bakdash. 2010. "The Effects of Acoustic Turn-by-Turn Navigation on Wayfinding." *Proceedings of the Human Factors and Ergonomics Society: Annual Meeting* 54:1926–1930.

Fernandino, L., L. L. Conant, J. R. Binder, K. Blindauer, B. Hiner, K. Spangler, and R. H. Desai. 2013. "Parkinson's Disease Disrupts Both Automatic and Controlled Processing of Action Verbs." *Brain and Language* 127:65–74.

Fino, E., M. Menegatti, A. Avenanti, and M. Rubini. 2016. "Enjoying vs. Smiling: Facial Muscular Activation in Response to Emotional Language." *Biological Psychology* 118:126–135.

Finzi, E., and N. E. Rosenthal. 2014. "Treatment of Depression with OnabotulinumtoxinA: A Randomized, Double-Blind, Placebo Controlled Trial." *Journal of Psychiatric Research* 52:1–6.

Fiske, S. T., A. Cuddy, and P. Glick. 2007. "Universal Dimensions of Social Cognition: Warmth, then Competence." *Trends in Cognitive Science* 11:77–83.

Fitzpatrick, P., R. Diorio, M. J. Richardson, and R. C. Schmidt. 2013. "Dynamical Methods for Evaluating the Time-Dependent Unfolding of Social Coordination in Children with Autism." *Frontiers in Integrative Neuroscience* 7:21.

Fitzpatrick, P., J. A. Frazier, D. M. Cochran, T. Mitchell, C. Coleman, and R. C. Schmidt. 2016. "Impairments of Social Motor Synchrony Evident in Autism Spectrum Disorder." *Frontiers in Psychology* 7:1323.

Fitzpatrick, P., V. Romero, J. L. Amaral, A. Duncan, H. Barnard, M. J. Richardson, and R. C. Schmidt. 2017. "Social Motor Synchronization: Insights for Understanding Social Behavior in Autism." *Journal of Autism and Developmental Disorders* 47 (7): 2092–2107.

Flint, A. J., S. E. Black, I. Campbell-Taylor, G. F. Gailey, and C. Levinton. 1993. "Abnormal Speech Articulation, Psychomotor Retardation, and Subcortical Dysfunction in Major Depression." *Journal of Psychiatric Research* 27 (3): 309–319.

Flückiger, C., A. C. Del Re, B. E. Wampold, D. Symonds, and A. O. Horvath. 2012. *How Central Is the Alliance in Psychotherapy? A Multilevel Longitudinal Meta-analysis.* Washinghton, DC: American Psychological Association.

Fodor, J. A. 1975. *The Language of Thought.* New York: Thomas Y. Crowell.

Fodor, J. A. 1981. *Representations: Philosophical Essays on the Foundations of Cognitive Science.* Brighton, UK: Harvester Press.

Fodor, J. A. 1983. *The Modularity of Mind: An Essay on Faculty Psychology.* Cambridge, MA: MIT Press.

Fodor, J. A. 1987. *Psychosemantics: The Problem of Meaning in the Philosophy of Mind.* Vol. 2. Cambridge, MA: MIT Press.

Fodor, J. A. 1990. *A Theory of Content and Other Essays.* Cambridge, MA: MIT Press.

Fodor, J. A. 1994. *The Elm and the Expert.* Cambridge, MA: MIT Press.

Fodor, J. A. 2000. *The Mind Doesn't Work That Way.* Cambridge, MA: MIT Press.

Fodor, J. A. 2009. "Where Is My Mind?" *London Review of Books* 31 (3): 13–15.

Fodor, J. A., and Z. Pylyshyn. 1988. "Connectionism and Cognitive Architecture: A Critical Analysis." *Cognition* 28:3–71.

Fogassi, L., P. F. Ferrari, B. Gesierich, S. Rozzi, F. Chersi, and G. Rizzolatti. 2005. "Parietal Lobe: From Action Organization to Intention Understanding." *Science* 308 (5722): 662–667.

Foglia, L., and R. A. Wilson. 2013. "Embodied Cognition." *WIREs Cognitive Science* 4:319–325.

Foroni, F., and G. R. Semin. 2013. "Comprehension of Action Negation Involves Inhibitory Simulation." *Frontiers in Human Neuroscience* 7:209.

Fournier, K. A., C. J. Hass, S. K. Naik, N. Lodha, and J. H. Cauraugh. 2010. "Motor Coordination in Autism Spectrum Disorders: A Synthesis and Meta-analysis." *Journal of Autism and Developmental Disorders* 40:1227–1240.

Frank, E., G. B. Cassano, P. Rucci, W. K. Thompson, H. C. Kraemer, A. Fagiolini, et al. 2011. "Predictors and Moderators of Time to Remission of Major Depression with Interpersonal Psychotherapy and SSRI Pharmacotherapy." *Psychological Medicine* 41 (1): 151–162.

Freitag, C. M., C. Kleser, M., Schneider, and A. von Gontard. 2007. "Quantitative Assessment of Neuromotor Function in Adolescents with High Functioning Autism and Asperger Syndrome." *Journal of Autism and Developmental Disorders* 37: 948–959.

Friston, K. J., K. E. Stephan, R. Montague, and R. J. Dolan. 2014. "Computational Psychiatry: The Brain as a Phantastic Organ." *Lancet Psychiatry* 1 (2): 148–158.

Frith, C. D. 2004. "Schizophrenia and Theory of Mind." *Psychological Medicine* 34 (3): 385–389.

Frith, U. 2012. "Why We Need Cognitive Explanations of Autism." *Quarterly Journal of Experimental Psychology* 65 (11): 2073–2092.

Fuchs, T. 2002. "The Phenomenology of Shame, Guilt and the Body in Body Dysmorphic Disorder and Depression." *Journal of Phenomenological Psychology* 33 (2): 223–243.

Fuchs, T. 2005. "Corporealized and Disembodied Minds: A Phenomenological View of the Body in Melancholia and Schizophrenia." *Philosophy, Psychiatry and Psychology* 12:95–107.

Fuchs, T., and H. De Jaegher. 2009. "Enactive Intersubjectivity: Participatory Sense-Making and Mutual Incorporation." *Phenomenology and the Cognitive Sciences* 8 (4): 465–486.

Fuchs, T., and J. E. Schlimme. 2009. "Embodiment and Psychopathology: A Phenomenological Perspective." *Current Opinion in Psychiatry* 22 (6): 570–575.

Fulford, K. W. M. 2000. "Philosophy Meets Psychiatry in the Twentieth Century—Four Looks Back and a Brief Look Forward. In *Philosophy Meets Medicine*, edited by P. Louhaiala and S. Stenman, 114–131. Helsinki: Helsinki University Press.

Fulford, K. W. M., K. Morris, J. Z. Sadler., and G. Stanghellini. 2003. "Past Improbable, Future Possible: The Renaissance in Philosophy and Psychiatry." In *Nature and Narrative: An Introduction to the New Philosophy of Psychiatry*, edited by K. W. M. Fulford, K. J. Morris, J. Z. Sadler, and G. Stanghellini, 1–41. Oxford: Oxford University Press.

Galbusera, L., M. T. Finn, and T. Fuchs. 2016. "Interactional Synchrony and Negative Symptoms: An Outcome Study of Body-Oriented Psychotherapy for Schizophrenia." *Psychotherapy Research* 28 (3): 457–469.

Gallagher, S. 2001a. "Empathy Simulation Narrative." *Journal of Consciousness Studies* 8 (5–7): 33–50.

Gallagher, S. 2001b. "The Practice of Mind: Theory, Simulation or Primary Interaction?" *Journal of Consciousness Studies* 8 (5–6): 83–108.

Gallagher, S. 2005. *How the Body Shapes the Mind*. Oxford: Oxford University Press.

Gallagher, S. 2007. "Simulation Trouble." *Social Neuroscience* 2 (3–4): 353–365.

Gallagher, S. 2008. "Inference or Interaction: Social Cognition without Precursors." *Philosophical Explorations* 11 (3): 163–174.

Gallagher, S. 2013a. "Intersubjectivity and Psychopathology." In *The Oxford Handbook of Philosophy and Psychiatry*, edited by K. W. M. Fulford, M. Davies, R. G. T. Gipps, G. Graham, J. Z. Sadler, G. Stanghellini, and T. Thornton, 258–274. Oxford: Oxford University Press.

Gallagher, S. 2013b. "The Socially Extended Mind." *Cognitive Systems Research* 25/26:4–12.

Gallagher, S. 2014. "Phenomenology and Embodied Cognition." In *The Routledge Handbook of Embodied Cognition*, edited by L. Shapiro, 9–18. London: Routledge.

Gallagher, S. 2017. *Enactivist Interventions: Rethinking the Mind*. Oxford: Oxford University Press.

Gallagher, S., and F. J. Varela. 2003. "Redrawing the Map and Resetting the Time: Phenomenology and the Cognitive Sciences." *Canadian Journal of Philosophy* 33 (supp. 1): 93–132.

Gallagher, S., and S. Varga. 2015a. "Conceptual Issues in Autism Spectrum Disorders." *Current Opinion in Psychiatry* 28 (2): 127–132.

Gallagher, S., and S. Varga. 2015b. "Social Cognition and Psychopathology." *World Psychiatry* 14 (1): 5–14.

Gallagher, S., and D. Zahavi. 2010. *The Phenomenological Mind: An Introduction to Philosophy of Mind and Cognitive Science*. Reprint. London: Routledge.

Gallese, V. 2005. "Embodied Simulation: From Neurons to Phenomenal Experience." *Phenomenology and the Cognitive Sciences* 4:23–48.

Gallese, V. 2006. "Intentional Attunement: A Neurophysiological Perspective on Social Cognition and Its Disruption in Autism." *Brain Research* 1079:15–24.

Gallese, V. 2007. "Before and Below 'Theory of Mind': Embodied Simulation and the Neural Correlates of Social Cognition." *Philosophical Transactions of the Royal Society of London B: Biological Sciences* 362 (1480): 659–669.

Gallese, V. 2011. "Embodied Simulation Theory: Imagination and Narrative." *Neuropsychoanalysis* 13 (2): 196–200.

Gallese, V. 2014. "Bodily Selves in Relation: Embodied Simulation as Second-Person Perspective on Intersubjectivity." *Philosophical Transactions of the Royal Society of London B: Biological Sciences* 369 (1644): 20130177. https://doi.org/10.1098/rstb.2013.0177.

Gallese, V., L. Fadiga, L. Fogassi, and G. Rizzolatti. 1996. "Action Recognition in the Premotor Cortex." *Brain* 119 (2): 593–609.

Gallese, V., and A. Goldman. 1998. "Mirror Neurons and the Simulation Theory of Mind-Reading." *Trends in Cognitive Sciences* 2 (12): 493–501.

Gallese, V., and G. Lakoff. 2005. "The Brain's Concepts: The Role of the Sensory-Motor System in Conceptual Knowledge." *Cognitive Neuropsychology* 22 (3–4): 455–479.

Gallese, V., M. J. Rochat, and C. Berchio. 2013. "The Mirror Mechanism and Its Potential Role in Autism Spectrum Disorder." *Developmental Medicine and Child Neurology* 55 (1): 15–22.

Gallese, V., M. J. Rochat, G. Cossu, and C. Sinigaglia. 2009. "Motor Cognition and Its Role in the Phylogeny and Ontogeny of Action Understanding." *Developmental Psychology* 45 (1): 103–113.

Garrod, S., and M. J. Pickering. 2009. "Joint Action, Interactive Alignment, and Dialog." *Topics in Cognitive Science* 1 (2): 292–304.

Gibbs, R. W., Jr. 2006. *Embodiment and Cognitive Science.* New York: Cambridge University Press

Gibbs, R. W., Jr. 2011. "Evaluating Conceptual Metaphor Theory." *Discourse Processes* 48 (8): 529–562.

Gibson, J. J. 1979. *The Ecological Approach to Visual Perception.* London: Erlbaum.

Giessner, S. R., and T. S. T. Schubert. 2007. "High in the Hierarchy: How Vertical Location and Judgments of Leaders' Power are Interrelated." *Organizational Behavior and Human Decision Processes* 104 (1): 30–44.

Gigerenzer, G., and W. Gaissmaier. 2011. "Heuristic Decision Making." *Annual Review of Psychology* 62:451–482.

Gigerenzer, G., and P. M. Todd. 1999. *Simple Heuristics That Make Us Smart.* New York: Oxford University Press.

Gilson, M., and A. Freeman. 2009. *Overcoming Depression: A Cognitive Therapy Approach Therapist Guide*. Oxford: Oxford University Press.

Gjelsvik, B., D. Lovric, and G. J. Mark G. 2018. "Embodied Cognition and Emotional Disorders: Embodiment and Abstraction in Understanding Depression." *Journal of Experimental Psychopathology*. https://doi.org/10.5127/pr.035714.

Glazebrook, C., D. Gonzalez, S. Hansen, and D. Elliott. 2009. "The Role of Vision for Online Control of Manual Aiming Movements in Persons with Autism Spectrum Disorders." *Autism* 13 (4): 411–433.

Glenberg, A. M. 2010. "Embodiment as a Unifying Perspective for Psychology." *WIREs: Cognitive Science* 1 (4): 586–596.

Glenberg, A. M., D. Havas, R. Becker, and M. Rinck. 2005. "Grounding Language in Bodily States." In *Grounding Cognition: The Role of Perception and Action in Memory, Language, and Thinking*, edited by D. Pecher and R. A. Zwaan, 115–128. Cambridge: Cambridge University Press.

Glenberg, A. M., and M. P. Kaschak. 2002. "Grounding Language in Action." *Psychonomic Bulletin and Review* 9:558–565.

Glenberg, A. M., J. K. Witt, and J. Metcalfe. 2013. "From the Revolution to Embodiment: 25 Years of Cognitive Psychology." *Perspectives on Psychological Science* 8 (5): 573–585.

Gockel, C., P. M. Kolb, and L. Werth. 2014. "Murder or Not? Cold Temperature Makes Criminals Appear to Be Cold-Blooded and Warm Temperature to Be Hot-Headed." *PLOS ONE* 9 (4): e96231.

Godfrey-Smith, P. 2005. "Folk Psychology as a Model." *Philosophers' Imprint* 5 (6): 1–16.

Goldin-Meadow, S. 2003. *Hearing Gesture: How Our Hands Help Us Think*. Cambridge, MA: Harvard University Press.

Goldman, A. I. 1989. "Interpretation Psychologized." *Mind & Language* 4 (3): 161–185.

Goldman, A. I. 2000. "Folk Psychology and Mental Concepts." *Proto Sociology* 14:4–25.

Goldman, A. I. 2006. *Simulating Minds: The Philosophy, Psychology, and Neuroscience of Mindreading*. Oxford: Oxford University Press.

Goldman, A. I. 2012. "Theory of Mind." In *The Oxford Handbook of Philosophy of Cognitive Science*, edited by E. Margolis, R. Samuels, and S. P. Stich, 402–424. Oxford: Oxford University Press.

Goldman, A. I. 2014. "The Bodily Formats Approach to Embodied Cognition." In *Current Controversies in Philosophy of Mind*, edited by U. Kriegel, 91–108. London: Routledge.

Gollwitzer, M., and A. Melzer. 2012. "Macbeth and the Joystick: Evidence for Moral Cleansing after Playing a Violent Video Game." *Journal of Experimental Social Psychology* 48 (6): 1356–1360.

Gopnik, A., A. N. Meltzoff, and P. Bryant. 1997. *Words, Thoughts, and Theories*. Vol. 1. Cambridge, MA: MIT Press.

Gopnik, A., and H. M. Wellman. 1994. "The Theory Theory." In *Mapping the Mind: Domain Specificity in Cognition and Culture*, edited by L. A. Hirschfeld and S. A. Gelman, 257–293. New York: Cambridge University Press.

Gordon, R. M. 1986. "Folk Psychology as Simulation." *Mind & Language* 1 (2): 158–171.

Gould, S. J., and E. S. Vrba. 1991. "Exaptation: A Missing Term in the Science of Form." *Journal of Social Issues* 47:43–65.

Gowen, E., and A. Hamilton. 2013. "Motor Abilities in Autism: A Review Using a Computational Context." *Journal of Autism and Developmental Disorders* 43 (2): 323–344.

Graham, G. 2009. *The Disordered Mind: An Introduction to Philosophy of Mind and Mental Illness*. London: Routledge.

Graham, G., and T. Horgan. 1994. "Southern Fundamentalism and the End of Philosophy." *Philosophical Issues* 5:219–247.

Graham, G., and G. L. Stephens, eds. 1994. *Philosophical Psychopathology*. Cambridge, MA: MIT Press.

Grant, A., M. Townend, R. Mulhern, and N. Short. 2010. *Cognitive Behavioural Therapy in Mental Health Care*. London: Sage.

Grèzes, J., B. Wicker, S. Berthoz, and B. De Gelder. 2009. "A Failure to Grasp the Affective Meaning of Actions in Autism Spectrum Disorder Subjects." *Neuropsychologia* 47 (8): 1816–1825.

Griffiths, P. E., and K. Stotz. 2006. "Genes in the Postgenomic Era." *Theoretical Medicine and Bioethics* 27 (6): 499–521.

Gross, J. J. 2014. "Emotion Regulation: Conceptual and Empirical Foundations." In *Handbook of Emotion Regulation*, 2nd ed., edited by J. J. Gross, 3–20. New York: Guilford Press.

Gross, J. J., and L. F. Barrett. 2011. "Emotion Generation and Emotion Regulation: One or Two Depends on Your Point of View." *Emotion Review* 3 (1): 8–16.

Gross, J. J., J. M. Richards, and O. P. John. 2006. "Emotion Regulation in Everyday Life." In *Emotion Regulation in Families: Pathways to Dysfunction and Health*, edited by D. K. Snyder, J. A. Simpson, and J. N. Hughes, 13–35. Washington, DC: American Psychological Association.

Gross, J. J., G. Sheppes, and H. L. Urry. 2011. "Emotion Generation and Emotion Regulation: A Distinction We Should Make (Carefully)." *Cognition and Emotion* 25:765–781.

Gross, J. J., and R. A. Thompson. 2007. "Emotion Regulation: Conceptual Foundations." In *Handbook of Emotion Regulation*, edited by J. J. Gross, 3–24. New York: Guilford Press.

Grossberg, S., and D. Seidman. 2006. "Neural Dynamics of Autistic Behaviors: Cognitive, Emotional, and Timing Substrates." *Psychological Review* 113 (3): 483–525.

Grube, M. 2006. "Towards an Empirically Based Validation of Intuitive Diagnostic: Rumke's 'Praecox Feeling' across the Schizophrenia Spectrum: Preliminary Results." *Psychopathology* 39:209–217.

Grush, R. 2002. "Cognitive Science." In *The Blackwell Guide to the Philosophy of Science*, edited by P. Machamer and M. Silberstein. https://doi.org/10.1002/9780470756614.ch13.

Guillaume, S., F. Jollant, I. Jaussent, N. Lawrence, A. Malafosse, and P. Courtet. 2009. "Somatic Markers and Explicit Knowledge Are Both Involved in Decision-Making." *Neuropsychologia* 47 (10): 2120–2124.

Gulsrud, A. C., L. B. Jahromi, and C. Kasari. 2010. "The Co-regulation of Emotions between Mothers and Their Children with Autism." *Journal of Autism and Developmental Disorders* 40 (2): 227–237.

Guo, X., Y. Park, N. D. Freedman, R. Sinha, A. R. Hollenbeck, A. Blair, and H. Chen. 2014. "Sweetened Beverages, Coffee, and Tea and Depression Risk among Older US Adults." *PLOS ONE* 9 (4): e94715. https://doi: 10.1371/journal.pone.0094715.

Guo, Y., D. R. Garfin, A. Ly, and W. A. Goldberg. 2017. "Emotion Coregulation in Mother-Child Dyads: A Dynamic Systems Analysis of Children with and without Autism Spectrum Disorder." *Journal of Abnormal Child Psychology* 45 (7): 1369–1383.

Häfner, M. 2013. "When Body and Mind Are Talking: Interoception Moderates Embodied Cognition." *Journal of Experimental Psychology* 60 (4): 255–259.

Haidt, J. 2007. "The New Synthesis in Moral Psychology." *Science* 316 (5827): 998–1002.

Hamilton, A. F. de C. 2013. "Reflecting on the Mirror Neuron System in Autism: A Systematic Review of Current Theories." *Developmental Cognitive Neuroscience* 3:91–105.

Happé, F. G. 1994. "An Advanced Test of Theory of Mind: Understanding of Story Characters' Thoughts and Feelings by Able Autistic, Mentally Handicapped, and Normal Children and Adults." *Journal of Autism and Developmental Disorders* 24 (2): 129–154.

Happé, F. G. 1995. *Autism: An Introduction to Psychological Theory*. Cambridge, MA: Harvard University Press.

Harari, H., S. G. Shamay-Tsoory, M. Ravid, and Y. Levkovitz. 2010. "Double Dissociation between Cognitive and Affective Empathy in Borderline Personality Disorder." *Psychiatry Research* 175:277–279.

Harinen, T. 2014. "Mutual Manipulability and Causal Inbetweenness." *Synthese* 195 (1): 35–54.

Harris, C. B., A. J. Barnier, J. Sutton, and P. G. Keil. 2014. "Couples as Socially Distributed Cognitive Systems: Remembering in Everyday Social and Material Contexts." *Memory Studies* 7 (3): 285–297.

Harrist, A. W., and R. M. Waugh. 2002. "Dyadic Synchrony: Its Structure and Function in Children's Development." *Developmental Review* 22:555–592.

Hatfield, E., J. T. Cacioppo, and R. L. Rapson. 1994. *Emotional Contagion: Cambridge Studies in Emotion and Social Interaction*. Cambridge: Cambridge University Press.

Hatfield, E., R. L. Rapson, and Y. L. Le. 2009. "Emotional Contagion and Empathy." In *The Social Neuroscience of Empathy*, edited by J. Decety and W. Ickes, 19–30. Cambridge, MA: MIT Press.

Haugeland, J. 1995. "Mind Embodied and Embedded." In *Mind and Cognition: Philosophical Perspectives on Cognitive Science and Artificial Intelligence*, Acta Philosophical Fennica, vol. 58, edited by L. Haaparanta and S. Heinämaa, 233–267. Helsinki: Finnish Philosophical Society.

Haugeland, J. 2000. *Having Thought*. Cambridge, MA: Harvard University Press.

Hauk, O., I. Johnsrude, and F. Pulvermüller. 2004. "Somatotopic Representation of Action Words in Human Motor and Premotor Cortex." *Neuron* 41:301–307.

Havas, D. A., A. M. Glenberg, and M. Rinck. 2007. "Emotion Simulation during Language Comprehension." *Psychonomic Bulletin and Review* 14 (3): 436–441.

Heal, J. 1996. "Simulation, Theory, and Content." In *Theories of Theories of Mind*, edited by P. Carruthers and P. K. Smith, 75–89. Cambridge: Cambridge University Press.

Heidegger, M. 1962. *Being and Time*. Translated by J. Macquarrie and E. Robinson. New York: Harper and Row.

Heimann, M. 2002. "Notes on Individual Differences and the Assumed Elusiveness of Neonatal Imitation." In *The Imitative Mind: Development, Evolution, and Brain Bases*, edited by A. N. Meltzoff and W. Prinz, 74–84. Cambridge: Cambridge University Press.

Henkel, L. A. 2014. "Point-and-Shoot Memories: The Influence of Taking Photos on Memory for a Museum Tour." *Psychological Science* 25:396–402.

Heyes, C. M., and C. D. Frith. 2014. "The Cultural Evolution of Mind Reading." *Science* 344 (6190): 1243091. https://doi.org/10.1126/science.1243091.

Hinton, G. E. 1992. "How Neural Networks Learn from Experience." *Scientific American* 267 (3): 144–151.

Hirschler-Guttenberg, Y., R. Feldman, S. Ostfeld-Etzion, N. Laor, and O. Golan. 2015. "Self- and Co-regulation of Anger and Fear in Preschoolers with Autism Spectrum Disorders: The Role of Maternal Parenting Style and Temperament." *Journal of Autism and Developmental Disorders* 45 (9): 3004–3014.

Hoffman, G. 2012. "Embodied Cognition for Autonomous Interactive Robots." *Topics in Cognitive Science* 4:759–772.

Holland, R. S. T., M. Hendriks, and S. T. Aarts. 2005. "Smells like Clean Spirit: Nonconscious Effects of Scent on Cognition and Behavior." *Psychological Science* 16:689–693.

Hollon, S. D., and A. T. Beck. 2004. "Cognitive and Cognitive-Behavioral Therapies." In *Bergin and Garfield's Handbook of Psychotherapy and Behavior Change: An Empirical Analysis*, 5th ed., edited by M. J. Lambert, 447–492. New York: Wiley.

Hollon, S. D., M. O. Stewart, and D. Strunk. 2006. "Enduring Effects for Cognitive Behavior Therapy in the Treatment of Depression and Anxiety." *Annual Review of Psychology* 57:285–315.

Hoogendoorn, W. E., M. N. van Poppel, P. M. Bongers, B. W. Koes, and L. M. Bouter. 2000. "Systematic Review of Psychosocial Factors at Work and Private Life as Risk Factors for Back Pain." *Spine* 25 (16): 2114–2125.

Horgan, T., and U. Kriegel. 2008. "Phenomenal Intentionality Meets the Extended Mind." *Monist* 91 (2): 347–373.

Horgan, T., and J. Tienson. 1991. "Settling into a New Paradigm." In *Connectionism and the Philosophy of Mind*, edited by T. Horgan and J. Tienson, 241-260. Dordrecht: Kluwer Academic Publishers.

Horgan, T., and J. Tienson. 1992a. "Cognitive Systems as Dynamical Systems." *Topoi* 11 (1): 27–43.

Horgan, T., and J. Tienson. 1992b. "Structured Representations in Connectionist Systems?" In *Connectionism: Theory and Practice*, edited by S. Davis, 195–228. Oxford: Oxford University Press.

Horst, S. 2007. *Beyond Reduction: Philosophy of Mind and Post-reductionist Philosophy of Science*. Oxford: Oxford University Press.

Horvath A., A. C. Del Re, C. Flückiger, and D. Symonds. 2011. "Alliance in Individual Psychotherapy." In *Psychotherapy Relationships That Work*, 2nd ed., edited by J. C. Norcross, 25–69. New York: Oxford University Press.

Hove, M. J., and J. L. Risen. 2009. "It's All in the Timing: Interpersonal Synchrony Increases Affiliation." *Social Cognition* 27 (6): 949–960.

Hubert, B. E., B. Wicker, E. Monfardini, and C. Deruelle. 2009. "Electrodermal Reactivity to Emotion Processing in Adults with Autistic Spectrum Disorders." *Autism* 13 (1): 9–19.

Huebner, B. 2014. *Macrocognition: A Theory of Distributed Minds and Collective Intentionality*. Oxford: Oxford University Press.

Humphrey, N. K. 1976. "The Social Function of Intellect." In *Growing Points in Ethology*, edited by P. P. G. Bateson and R. A. Hinde, 303–317. Cambridge: Cambridge University Press.

Humphrey, N., and D. C. Dennett. 1989. "Speaking for Ourselves." *Raritan* 9:68–98.

Hunt, E. 1987. "A Cognitive Model of Individual Differences, with an Application to Attention." In *Intelligence and Cognition*, edited by S. H. Irvine and S. E. Newstead, 177–215. Dodrecht: Nijhoff.

Hurley, S. 1998. *Consciousness in Action*. Cambridge, MA: Harvard University Press.

Hurley, S. 2008. "The Shared Circuits Model (SCM): How Control, Mirroring, and Simulation Can Enable Imitation, Deliberation, and Mindreading." *Behavioral and Brain Sciences* 31 (1): 1–22.

Hurley, S. 2010. "Varieties of Externalism." In *The Extended Mind*, edited by R. Menary, 101–153. Cambridge, MA: MIT Press.

Hutchins, E. 1995a. *Cognition in the Wild*. Cambridge, MA: MIT Press.

Hutchins, E. 1995b. "How a Cockpit Remembers Its Speeds." *Cognitive Science* 19:265–288.

Hutchins, E. 2011. "Enculturating the Supersized Mind." *Philosophical Studies* 152 (3): 437–446.

Hutchins E. 2014. "The Cultural Ecosystem of Human Cognition." *Philosophical Psychology* 27:34–49.

Hutto, D. D., M. Herschbach, and V. Southgate. 2011. "Editorial: Social Cognition: Mindreading and Alternatives." *Review of Philosophy and Psychology* 2 (3): 375–395.

Hutto, D. D., M. D. Kirchhoff, and E. Myin. 2014. "Extensive Enactivism: Why Keep It All In?" *Frontiers in Human Neuroscience* 8:706.

Hutto, D. D., and E. Myin. 2013. *Radicalizing Enactivism: Basic Minds without Content.* Cambridge, MA: MIT Press.

Huys, Q. J. M., M. Moutoussis, and J. Williams. 2011. "Are Computational Models of Any Use to Psychiatry?" *Neural Networks* 24 (6): 544–551.

Iacoboni, M., and M. Dapretto. 2006. "The Mirror Neuron System and the Consequences of Its Dysfunction." *Nature Reviews Neuroscience* 7 (12): 942–951.

Iacoboni, M., I. Molnar-Szakacs, V. Gallese, G. Buccino, J. C. Mazziotta, and G. Rizzolatti. 2005. "Grasping the Intentions of Others with One's Own Mirror Neuron System." *PLOS Biology* 3 (3): 0529–0535.

Iao, L.-S., and S. R. Leekam. 2014. "Nonspecificity and Theory of Mind: New Evidence from a Nonverbal False-Sign Task and Children with Autism Spectrum Disorders." *Journal of Experimental Child Psychology* 122:1–20.

IJzerman, S. T., M. Gallucci, S. T. T. J. L. Pouw, S. C. Weiβgerber, N. J. Van Doesum, and K. D. Williams. 2012. "Cold-Blooded Loneliness: Social Exclusion Leads to Lower Skin Temperatures." *Acta Psychologica* 140:283–288.

IJzerman, S. T., and G. R. Semin. 2010. "Temperature Perceptions as a Ground for Social Proximity." *Journal of Experimental Social Psychology* 46:867–873.

Illari, P. M., and J. Williamson. 2012. "What Is a Mechanism? Thinking about Mechanisms across the Sciences." *European Journal for Philosophy of Science* 2 (1): 119–135.

Ingram, R. E. 1984. "Toward an Information-Processing Analysis of Depression." *Cognitive Therapy and Research* 8 (5): 443–477.

Ingram, R. E., G. J. Siegle, and D. Steidtmann. 2009. "Methodological Issues in the Study of Depression." In *Handbook of Depression*, edited by I. H. Gotlib and C. L. Hammen, 69–92. New York: Guilford Press.

Isomura, T., and T. Nakano. 2016. "Automatic Facial Mimicry in Response to Dynamic Emotional Stimuli in Five-Month-Old Infants." *Proceedings of the Royal Society of London B: Biological Sciences* 283:20161948.

Jarrold, C., P. Smith, J. Boucher, and P. Harris. 1994. "Comprehension of Pretense in Children with Autism." *Journal of Autism and Developmental Disorders* 24: 433–455.

Johnson, C. 1999. "Metaphor vs. Conflation in the Acquisition of Polysemy: The Case of SEE." In *Cultural, Psychological and Typological Issues in Cognitive Linguistics: Selected Papers of the Bi-annual ICLA Meeting in Albuquerque, July 1995*, edited by M. K. Hiraga, C. Sinha, and S. Wilcox, 155–170. Amsterdam: John Benjamins Publishing.

Joseph, R. M., H, Tager-Flusberg, and C. Lord. 2002. "Cognitive Profiles and Social-Communicative Functioning in Children with Autism Spectrum Disorder." *Journal of Child Psychology and Psychiatry, and Allied Disciplines* 4 (6): 807–821.

Jostmann, N., D. Lakens, and T. Schubert. 2009. "Weight as an Embodiment of Importance." *Psychological Science* 20:1169–1174.

Kaiser, M. I., and B. Krickel. 2016. "The Metaphysics of Constitutive Mechanistic Phenomena." *British Journal for the Philosophy of Science* 3 (1):745–779.

Kang, S. T., L. Williams, M. Clark, J. Gray, and J. Bargh. 2010. "Physical Temperature Effects on Trust Behavior: The Role of Insula." *Social Cognitive and Affective Neuroscience* 6 (4): 507–515.

Kanner, L. 1943. "Autistic Disturbances of Affective Contact." *Nervous Child* 2:217–250.

Kaplan, D. M. 2012. "How to Demarcate the Boundaries of Cognition." *Biology & Philosophy* 27 (4): 545–570.

Kaspar, K., V. Krapp, and P. König. 2015. "Hand Washing Induces a Clean Slate Effect in Moral Judgments: A Pupillometry and Eye-Tracking Study." *Scientific Reports* 5:10471.

Kaspar, K., and A. Vennekötter. 2015. "Embodied Information in Cognitive Tasks: Haptic Weight Sensations Affect Task Performance and Processing Style." *Advances in Cognitive Psychology* 11 (3): 64–76.

Kassel, J. D., M. Bornovalova, and N. Mehta. 2007. "Generalized Expectancies for Negative Mood Regulation Predict Change in Anxiety and Depression among College Students." *Behaviour Research and Therapy* 45 (5): 939–950.

Kästner, L. 2017. *Philosophy of Cognitive Neuroscience: Causal Explanations, Mechanisms and Experimental Manipulations*. Berlin: De Gruyter.

Kauppinen, A. 2007. "The Rise and Fall of Experimental Philosophy." *Philosophical Explorations* 10 (2): 95–118.

Kelso, J. A. 2009. "Coordination Dynamics." In *Encyclopedia of Complexity and Systems Science*, edited by R. A. Meyers, 1537–1565. New York: Springer.

Kelso, S. 1995. *Dynamic Patterns*. Cambridge, MA: MIT Press.

Kendler, K. S. 2008. "Explanatory Models for Psychiatric Illness." *American Journal of Psychiatry* 165 (6): 695–702.

Kendler, K. S., and J. Campbell. 2009. "Interventionist Causal Models in Psychiatry: Repositioning the Mind–Body Problem." *Psychological Medicine* 39 (6): 881–887.

Kendler, K. S., P. Zachar, and C. Craver. 2011. "What Kinds of Things Are Psychiatric Disorders?" *Psychological Medicine* 41 (6): 1143–1150.

Kerr, N., R. I. M. Dunbar, and R. Bentall. 2003. "Theory of Mind Deficits in Bipolar Affective Disorder." *Journal of Affective Disorders* 73:253–259.

Keysers, C., J. H. Kaas, and V. Gazzola. 2010. "Somatosensation in Social Perception." *Nature Reviews Neuroscience* 11 (6): 417–428.

Keysers, C., M. Thioux, and V. Gazzola. 2013. "Mirror Neuron System and Social Cognition." In *Understanding Other Minds: Perspectives from Developmental Social Neuroscience*, edited by S. Baron-Cohen, M. Lombardo, H. Tager-Flusberg, and D. Cohen, 223–256. Oxford: Oxford University Press.

Keysers, C., B. Wicker, V. Gazzola, J.-L. Anton, L. Fogassi, and V. Gallese. 2004. "A Touching Sight: SII/PV Activation during the Observation and Experience of Touch." *Neuron* 42 (2): 335–346.

Kilner, J. M., A. Neal, N. Weiskopf, K. J. Friston, and C. D. Frith. 2009. "Evidence of Mirror Neurons in Human Inferior Frontal Gyrus." *Journal of Neuroscience* 29 (32): 10153–10159.

Kilner, J. M., C. Vargas, S. Duval, S.-J. Blakemore, and A. Sirigu. 2004. "Motor Activation Prior to Observation of a Predicted Movement." *Nature Neuroscience* 7: 1299–1301. http://dx.doi.org/ 10.1038/nn1355.

Kim, J. 1993. "Mental Causation in a Physical World." *Philosophical Issues* 3:157–176.

Kirchhoff, M. D. 2015. "Extended Cognition and the Alleged Causal-Constitutive Fallacy: In Search for a Diachronic and Dynamical Conception of Constitution." *Philosophy and Phenomenological Research* 40 (2): 320–360.

Kirchhoff, M. D. 2014. "Extended Cognition and Constitution: Re-evaluating the Constitutive Claim of Extended Cognition." *Philosophical Psychology* 27 (2): 258–283.

Kirchhoff, M. D. 2017. "From Mutual Manipulation to Cognitive Extension: Challenges and Implications." *Phenomenology and the Cognitive Sciences* 16 (5): 863–878.

Kirschner, S., and M. Tomasello. 2010. "Joint Music Making Promotes Prosocial Behavior in 4-Year-Old Children." *Evolution and Human Behavior* 31 (5): 354–364.

Kistler, M. 2010. "Causation across Levels, Constitution, and Constraint." In *EPSA Philosophical Issues in the Sciences*, edited by M. Suárez, M. Dorato, and M. Rédei, 141–151. Dordrecht: Springer.

Koehne, S., A. Hatri, J. T. Cacippo, and I. Dziobek. 2016. "Perceived Interpersonal Synchrony Increases Empathy: Insights from Autism Spectrum Disorder." *Cognition* 146:8–15.

Kolak, D., ed. 2006. *Cognitive Science: An Introduction to Mind and Brain*. London: Routledge.

Koole, S. L., and W. Tschacher. 2016. "Synchrony in Psychotherapy: A Review and an Integrative Framework for the Therapeutic Alliance." *Frontiers in Psychology* 7: 862.

Kornblith, H. 2017. "A Naturalistic Methodology." In *The Cambridge Companion to Philosophical Methodology*, edited by G. D'Oro and S. Overgaard, 150–172. Cambridge: Cambridge University Press.

Körner, A., S. Topolinski, and F. Strack. 2015. "Routes to Embodiment." *Frontiers in Psychology* 6:940.

Kouchaki, M., F. Gino, and A. Jami. 2014. "The Burden of Guilt: Heavy Backpacks, Light Snacks, and Enhanced Morality." *Journal of Experimental Psychology: General* 143:414–424.

Kövecses, Z. 2000. *Metaphor and Emotion*. New York: Cambridge University Press.

Kövecses, Z. 2005. *Metaphor in Culture: Universality and Variation*. Cambridge: Cambridge University Press.

Krebs, J. R., and R. Dawkins. 1984. "Animal Signals: Mind-Reading and Manipulation." In *Behavioural Ecology—An Evolutionary Approach*, 2nd ed., edited by J. R. Krebs and N. B. Davies, 380–405. Malden, MA: Blackwell Scientific Publications.

Krickel, B. 2017. "Making Sense of Interlevel Causation in Mechanisms from a Metaphysical Perspective." *Journal for General Philosophy of Science* 48 (3): 453–468.

Krickel, B. 2018. "Saving the Mutual Manipulability Account of Constitutive Relevance." *Studies in History and Philosophy of Science Part A* 68:58–67.

Kriegel, U. 2014. "Two Notions of Mental Representation." In *Current Controversies in Philosophy of Mind*, edited by U. Kriegel, 161–179. New York: Routledge.

Kumar, P., G. Waiter, T. Ahearn, M. Milders, I. Reid, and J. D. Steele. 2008. "Abnormal Temporal Difference Reward-Learning Signals in Major Depression." *Brain* 131 (8): 2084–2093.

Kupper, Z., F. Ramseyer, H. Hoffmann, S. Kalbermatten, and W. Tschacher. 2010. "Video-Based Quantification of Body Movement during Social Interaction Indicates the Severity of Negative Symptoms in Patients with Schizophrenia." *Schizophrenia Research* 121:90–100.

Kupper, Z., F. Ramseyer, H. Hoffmann, and W. Tschacher. 2015. "Nonverbal Synchrony in Social Interactions of Patients with Schizophrenia Indicates Socio-communicative Deficits." *PLOS ONE* 10 (12): e0145882.

Lacey, S., R. Stilla, and K. Sathian. 2012. "Metaphorically Feeling: Comprehending Textural Metaphors Activates Somatosensory Cortex." *Brain and Language* 120:416–421.

Lakens, D. 2010. "Movement Synchrony and Perceived Entitativity." *Journal of Experimental Social Psychology* 46 (5): 701–708.

Lakens, D. 2014. "Grounding Social Embodiment." *Social Cognition* 32:168–183.

Lakin, J. L., and T. L. Chartrand. 2012. "Behavioral Mimicry as an Affiliative Response to Social Exclusion." In *The Oxford Handbook of Social Exclusion*, edited by C. N. DeWall, 266–274. Oxford: Oxford University Press.

Lakoff, G. 2008. *The Political Mind: Why You Can't Understand 21st-Century Politics with an 18th-Century Brain*. New York: Viking.

Lakoff, G. 2012. "Explaining Embodied Cognition Results." *Topics in Cognitive Science* 4 (4): 773–785.

Lakoff, G. 2014. "Mapping the Brain's Metaphor Circuitry: Metaphorical Thought in Everyday Reason." *Frontiers in Human Neuroscience* 8:958.

Lakoff, G., and M. Johnson. 1999. *Philosophy in the Flesh: The Embodied Mind and its Challenge to Western Thought*. New York: Basic Books.

Lane, R. D., L. Nadel, J. J. Allen, and A. W. Kaszniak. 2000. "The Study of Emotion from the Perspective of Cognitive Neuroscience." In *Cognitive Neuroscience of Emotion*, edited by R. D. Lane and L. Nadel, 3–11. Oxford: Oxford University Press.

Larkin, F., S. Guerin, J. A. Hobson, and S. E. Gutstein. 2015. "The Relationship Development Assessment–Research Version: Preliminary Validation of a Clinical Tool and Coding Schemes to Measure Parent-Child Interaction in Autism." *Clinical Child Psychology and Psychiatry* 20 (2): 239–260.

Lavelle, M., P. G. Healey, and R. McCabe. 2013. "Is Nonverbal Communication Disrupted in Interactions Involving Patients with Schizophrenia?" *Schizophrenia Bulletin* 39:1150–1158.

Leahy, R. L. 2009. "Resistance: An Emotional Schema Therapy (EST) Approach." In *Cognitive Behaviour Therapy: A Guide for the Practising Clinician*, vol. 2, edited by G. Simos, 187–204. New York: Routledge

Leahy, R. L., D. Tirch, and L. A. Napolitano. 2011. *Emotion Regulation in Psychotherapy: A Practitioner's Guide*. New York: Guilford Press.

Leary, M. R., and D. A. Hill. 1996. "Moving On: Autism and Movement Disturbance." *Mental Retardation* 34:39–53.

Leclère, C., S. Viaux, M. Avril, C. Achard, M. Chetouani, S. Missonnier, and D. Cohen. 2014. "Why Synchrony Matters during Mother-Child Interactions: A Systematic Review." *PLoS ONE* 9 (12): e113571. https://doi.org/10.1371/journal.pone.0113571.

LeDoux, J. E. 1996. *The Emotional Brain*. New York: Simon and Schuster.

Lee, S. W., and N. Schwarz. 2010. "Dirty Hands and Dirty Mouths: Embodiment of the Moral-Purity Metaphor Is Specific to the Motor Modality Involved in Moral Transgression." *Psychological Science* 21 (10): 1423–1425.

Lee, S. W., and N. Schwarz. 2012. "Bidirectionality, Mediation, and Moderation of Metaphorical Effects: The Embodiment of Social Suspicion and Fishy Smells." *Journal of Personality and Social Psychology* 103:737–749.

Lee, S. W., S. T. Tang, J. Wan, X. Mai, and C. Liu. 2015. "A Cultural Look at Moral Purity: Wiping the Face Clean." *Frontiers in Psychology* 6:577. http://doi.org/10.3389/fpsyg.2015.00577.

Lemelin, S., and P. Baruch. 1998. "Clinical Psychomotor Retardation and Attention in Depression." *Journal of Psychiatric Research* 32 (2): 81–88.

Lemke, M. R., T. Wendorff, B. Mieth, K. Buhl, and M. Linnemann. 2000. "Spatiotemporal Gait Patterns during over Ground Locomotion in Major Depression Compared with Healthy Controls." *Journal of Psychiatric Research* 34 (4–5): 277–283.

Leslie, A. M. 1987. "Pretense and Representation: The Origins of 'Theory of Mind.'" *Psychological Review* 94 (4): 412–426.

Leslie, A. M. 1988. "Some Implications of Pretense for Mechanisms Underlying the Child's Theory of Mind." In *Developing Theories of Mind*, edited by J. W. Astington, P. L. Harris, and D. R. Olson, 19–46. New York: Cambridge University Press.

Leslie, A. M. 1992. "Pretense, Autism, and the Theory-of-Mind Module." *Current Directions in Psychological Science* 1 (1): 18–21.

Leslie, A. M. 1994. "Pretending and Believing: Issues in the Theory of ToMM." *Cognition* 50 (1): 211–238.

Leuridan, B. 2012. "Three Problems for the Mutual Manipulability Account of Constitutive Relevance in Mechanisms." *British Journal for the Philosophy of Science* 63 (2): 399–427.

Levitan, R., and J. Hirschberg. 2011. "Measuring Acoustic-Prosodic Entrainment with Respect to Multiple Levels and Dimensions." In *Twelfth Annual Conference of the International Speech Communication Association*, 3081–3084.

Lewis, D. 1980. "Mad Pain and Martian Pain." In *Readings in the Philosophy of Psychology*, edited by Ned Block, 216–222. Cambridge, MA: Harvard University Press.

Lewis, D. 1986. "Postscript c to 'Causation': (Insensitive Causation)." In *Philosophical Papers II*, 184–188. Oxford: Oxford University Press.

Lewis, V., and J. Boucher. 1988. "Spontaneous, Instructed and Elicited Play in Relatively Able Autistic Children." *British Journal of Developmental Psychology* 6:325–337.

Liljenquist, K., C. B. Zhong, and A. D. Galinsky. 2010. "The Smell of Virtue: Clean Scents Promote Reciprocity and Charity." *Psychological Science* 21 (3): 381–383.

Lindenberger, U., and U. Mayr. 2014. "Cognitive Aging: Is There a Dark Side to Environmental Support?" *Trends in Cognitive Sciences* 18 (1): 7–15.

Lobel, T. E., A. Cohen, L. Kalay Shahin, S. Malov, S. T. Golan, and S. Busnach. 2014. "Being Clean and Acting Dirty: The Paradoxical Effect of Self-Cleansing." *Ethics and Behavior* 25 (4): 307–313.

Lourenco, S. F., and M. R. Longo. 2011. "Origins and Development of Generalized Magnitude Representation." In *Space, Time and Number in the Brain*, edited by S. Dehaene and E. M. Brannon, 225–244. San Diego: Academic Press.

Louwerse, M. M., R. Dale, E. G. Bard, and P. Jeuniaux. 2012. "Behavior Matching in Multimodal Communication Is Synchronized." *Cognitive Science* 36 (8): 1404–1426.

Ludwig, D. 2015. *A Pluralist Theory of the Mind*. Vol. 2. Cham, Switzerland: Springer International Publishing.

Lumsden, J., L. K. Miles, and C. N. Macrae. 2014. "Sync or Sink? Interpersonal Synchrony Impacts Self-Esteem." *Frontiers in Psychology* 5:1064.

Lundqvist, L. O. 1995. "Facial EMG Reactions to Facial Expressions: A Case of Facial Emotional Contagion?" *Scandinavian Journal of Psychology* 36 (2): 130–141.

Macdonald, C., and G. Macdonald. 1995. *Connectionism: Debates on Psychological Explanation*. Malden, MA: Blackwell.

MacDonald, G., R. Kingsbury, and S. Shaw. 2005. "Adding Insult to Injury." In *The Social Outcast: Ostracism, Social Exclusion, Rejection, and Bullying*, edited by K. D. Williams, J. P. Forgas, and W. V. Hippel, 77–90. New York: Psychology Press.

MacDonald, G., and M. R. Leary. 2005. "Why Does Social Exclusion Hurt? The Relationship between Social and Physical Pain." *Psychological Bulletin* 131 (2): 202–223.

Machamer, P., L. Darden, and C. F. Craver. 2000. "Thinking about Mechanisms." *Philosophy of Science* 67 (1): 1–25.

Magid, M., J. S. Reichenberg, P. E. Poth, H. T. Robertson, A. K. LaViolette, T. H. Kruger, and M. Maiese. 2016. *Embodied Selves and Divided Minds*. Oxford: Oxford University Press.

Machery, E. 2009. *Doing without Concepts*. New York: Oxford University Press.

Maiese, M. 2016. *Embodied Selves and Divided Minds*. Oxford: Oxford University Press.

Manian, N., and M. H. Bornstein. 2009. "Dynamics of Emotion Regulation in Infants of Clinically Depressed and Nondepressed Mothers." *Journal of Child Psychology and Psychiatry* 50 (11): 1410–1418.

Mari, M., U. Castiello, D. Marks, C. Marraffa, and M. Prior. 2003. "The Reach-to-Grasp Movement in Children with Autism Spectrum Disorder." *Philosophical Transactions of the Royal Society of London B: Biological Sciences* 358 (1430): 393–403.

Marr, D. 1982. *Vision: A Computational Approach*. San Francisco: Freeman & Co.

Marroquín, B., and S. Nolen-Hoeksema. 2015. "Emotion Regulation and Depressive Symptoms: Close Relationships as Social Context and Influence." *Journal of Personality and Social Psychology* 109 (5): 836.

Marsh, K. L., R. W. Isenhower, M. J. Richardson, M. Helt, A. D. Verbalis, R. C. Schmidt, and D. Fein. 2013. "Autism and Social Disconnection in Interpersonal Rocking." *Frontiers in Integrative Neuroscience* 7. https://doi.org/10.3389/fnint.2013.00004.

Marsh, K. L., M. J. Richardson, and R. C. Schmidt. 2009. "Social Connection through Joint Action and Interpersonal Coordination." *Topics in Cognitive Science* 1 (2): 320–339.

Maurage, P., D. Grynberg, X. Noel, F. Joassin, P. Philippot, C. Hanak, P. Verbanck, O. Luminet, P. de Timary, and S. Campanella. 2011. "Dissociation between Affective and Cognitive Empathy in Alcoholism: A Specific Deficit for the Emotional Dimension." *Alcoholism, Clinical and Experimental Research* 35 (9) (September): 1662–1668. https://doi.org/10.1111/j.1530-0277.2011.01512.x.

Mazzurega, M., F. Pavani, M. P. Paladino, and T. W. Schubert. 2011. "Self-Other Bodily Merging in the Context of Synchronous but Arbitrary-Related Multisensory Inputs." *Experimental Brain Research* 213 (2–3): 213–221.

McBeath, M. K., D. M. Shaffer, and M. K. Kaiser. 1995. "How Baseball Outfielders Determine Where to Run to Catch Fly Balls." *Science* 268 (5210): 569–573.

McCulloch, W., and W. Pitts. 1943. "A Logical Calculus of the Ideas Immanent in Nervous Activity." *Bulletin of Mathematical Biophysics* 7:115–133.

McIntosh, D. N. 1996. "Facial Feedback Hypotheses: Evidence, Implications, and Directions." *Motivation and Emotion* 20 (2): 121–147.

McIntosh, D. N., A. Reichmann-Decker, P. Winkielman, and J. L. Wilbarger. 2006. "When the Social Mirror Breaks: Deficits in Automatic, but Not Voluntary, Mimicry of Emotional Facial Expressions in Autism." *Developmental Science* 9 (3): 295–302.

McLaughlin, B. P. 1993. "The Connectionism/Classicism Battle to Win Souls." *Philosophical Studies* 71:163–190.

McLaughlin, B. P. 2004. "Computationalism, Connectionism, and the Philosophy of Mind." In *The Blackwell Guide to the Philosophy of Computing and Information*, edited by L. Floridi, 135–152. Oxford: Blackwell Publishing.

McLaughlin, B. P. 2014. "Can an ICS Architecture Meet the Systematicity and Productivity Challenges?" In *Architecture of Cognition: Rethinking Fodor and Pylyshyn's Systematicity Challenge*, edited by P. Calvo and J. Symons, 31–76. Cambridge, MA: MIT Press.

McLaughlin, B. P., and K. Bennett. 2018. "Supervenience." In *The Stanford Encyclopedia of Philosophy*, edited by E. N. Zalta. https://plato.stanford.edu/archives/spr2018/entries/supervenience/.

McLeod, P., K. Plunkett, and E. T. Rolls. 1998. *Introduction to Connectionist Modelling of Cognitive Processes*. Oxford: Oxford University Press.

McMullen, L. M., and J. B. Conway. 2002. "Conventional Metaphors for Depression." In *The Verbal Communication of Emotions: Interdisciplinary Perspectives*, edited by S. R. Fussell, 167–181. New York: Psychology Press.

Meltzoff, A. N. 2005. "Imitation and Other Minds: The 'Like Me' Hypothesis." In *Perspectives on Imitation: From Neuroscience to Social Science*, vol. 2, edited by S. Hurley and N. Chater, 55–77. Cambridge, MA: MIT Press.

Meltzoff, A. N., and R. Brooks. 2001. "'Like Me' as a Building Block for Understanding Other Minds: Bodily Acts, Attention, and Intention." In *Intentions and Intentionality: Foundations of Social Cognition*, edited by B. F. Malle, L. J. Moses, and D. A. Baldwin, 171–191. Cambridge, MA: MIT Press.

Meltzoff, A. N., and M. K. Moore. 1983. "Newborn Infants Imitate Adult Facial Gestures." *Child Development* 54 (3): 702–709.

Meltzoff, A. N., and M. K. Moore. 1989. "Imitation in Newborn Infants: Exploring the Range of Gestures Imitated and the Underlying Mechanisms." *Developmental Psychology* 25 (6): 954–962.

Meltzoff, A. N., and M. K. Moore. 1995. "A Theory of the Role of Imitation in the Emergence of Self." In *The Self in Infancy: Theory and Research*, edited by P. Rochat, 73–93. Amsterdam: Elsevier.

Meltzoff, A. N., and M. K. Moore. 1997. "Explaining Facial Imitation: A Theoretical Model." *Early Development and Parenting* 6 (3–4): 179–192.

Menary, R. 2007. *Cognitive Integration: Mind and Cognition Unbounded*. Basingstoke, UK: Palgrave Macmillan.

Menary, R. 2010a. "Cognitive Integration and the Extended Mind." In *The Extended Mind*, edited by R. Menary, 227–244. Cambridge, MA: MIT Press.

Menary, R. 2010b. "Introduction: The Extended Mind in Focus." In *The Extended Mind*, edited by R. Menary, 1–26. Cambridge, MA: MIT Press.

Menary, R. 2010c. "Introduction to the Special Issue on 4E Cognition." *Phenomenology and the Cognitive Sciences* 9 (4): 459–463.

Menary, R. 2014. "Neural Plasticity, Neuronal Recycling and Niche Construction." *Mind & Language* 29 (3): 286–303.

Menary, R. 2015. "Mathematical Cognition: A Case of Enculturation." In *Open MIND*, edited by T. Metzinger and J. M. Windt. Frankfurt: MIND Group. https://doi.org/10.15502/9783958570115.

Mergl, R., G. Juckel, J. Rihl, V. Henkel, M. Karner, and P. Tigges. 2004. "Kinematical Analysis of Handwriting Movements in Depressed Patients." *Acta Psychiatrica Scandinavica* 109 (5): 383–391.

Merleau-Ponty, M. (1962) 2005. *Phenomenology of Perception*. Hoboken, NJ: Routledge.

Messerli, F. H. 2012. "Chocolate Consumption, Cognitive Function, and Nobel Laureates." *New England Journal of Medicine* 367 (16): 1562–1564.

Meyer-Lindenberg, A. 2008. "Trust Me on This." *Science* 321:778–780.

Michael, J. 2011. "Interactionism and Mindreading." *Review of Philosophy and Psychology* 2 (3): 559–578.

Michalak, J., J. Mischnat, and T. Teismann. 2014. "Sitting Posture Makes a Difference: Embodiment Effects on Depressive Memory Bias. *Clinical Psychology and Psychotherapy* 21 (6): 519–524. https://doi.org/10.1002/cpp.1890.

Michalak, J., K. Rohde, and N. F. Troje. 2015. "How We Walk Affects What We Remember: Gait Modifications through Biofeedback Change Negative Affective Memory Bias." *Journal of Behavior Therapy and Experimental Psychiatry* 46:121–125.

Michalak, J., N. F. Troje, J. Fischer, P. Vollmar, T. Heidenreich, and D. Schulte. 2009. "Embodiment of Sadness and Depression: Gait Patterns Associated with Dysphoric Mood." *Psychosomatic Medicine* 71 (5): 580–587.

Miles, L., J. L. Griffiths, M. J. Richardson, and C. N. Macrae. 2010. "Too Late to Coordinate: Contextual Influences on Behavioral Synchrony." *European Journal of Social Psychology* 40 (1): 52–60.

Millikan, R. 1984. *Language, Thought, and Other Biological Categories*. Cambridge, MA: MIT Press.

Millikan, R. 1989. "In Defense of Proper Functions." *Philosophy of Science* 56 (2): 288–302.

Millikan, R. 2004. *Varieties of Meaning*. Cambridge, MA: MIT Press.

Milner, A. D., and M. A. Goodale. 1995. *The Visual Brain in Action*. 2nd ed. Oxford: Oxford University Press.

Mitchell, S. D. 2009. *Unsimple Truths: Science, Complexity, and Policy*. London: University of Chicago Press.

Moody, E. J., and D. N. McIntosh. 2006. "Imitation in Autism Findings and Controversies." In *Imitation and the Social Mind*, edited by S. J. Rogers and J. H. G. Williams, 71–88. New York: Guilford Press.

Moody, E. J., and D. N. McIntosh. 2011. "Mimicry of Dynamic Emotional and Motor-Only Stimuli." *Social Psychological and Personality Science* 2 (6): 679–686.

Moore, G. A., and S. D. Calkins. 2004. "Infants' Vagal Regulation in the Still-Face Paradigm Is Related to Dyadic Coordination of Mother-Infant Interaction." *Developmental Psychology* 40 (6): 1068.

Mosconi, M. W., and J. A. Sweeney. 2015. "Sensorimotor Dysfunctions as Primary Features of Autism Spectrum Disorders." *Science China: Life Sciences* 58 (10): 1016–1023.

Mostofsky, S. H., P. Dubey, V. K. Jerath, E. M. Janciewicz, M. C. Goldberg, and M. B. Denckla. 2006. "Developmental Dyspraxia Is Not Limited to Imitation in Children with Autism Spectrum Disorders." *Journal of the International Neuropsychological Society* 12:314–326.

Mukamel, R., A. D. Ekstrom, J. Kaplan, M. Iacoboni, and I. Fried. 2010. "Single-Neuron Responses in Humans during Execution and Observation of Actions." *Current Biology* 20:750–756.

Murphy, D. 2008. "Concepts of Disease and Health." In *The Stanford Encyclopedia of Philosophy*, edited by E. N. Zalta. http://plato.stanford.edu/archives/sum2009/entries/health-disease/.

Murray, L., and C. Trevarthen. 1986. "The Infant's Role in Mother–Infant Communications." *Journal of Child Language* 13 (1): 15–29.

Nair, S., M. Sagar, J. Sollers III, N. Consedine, and E. Broadbent. 2015. "Do Slumped and Upright Postures Affect Stress Responses? A Randomized Trial." *Health Psychology* 34 (6): 632–641.

Nakamura, S. T., S. T. Ito, S. T. Honma, T. Mori, and J. Kawaguchi. 2014. "Cold-Hearted or Cool-Headed: Physical Coldness Promotes Utilitarian Moral Judgment." *Frontiers in Psychology* 5:1086.

Néda, Z., E. Ravasz, Y. Brechet, T. Vicsek, and A.-L. Barabási. 2000. "Self-Organizing Processes: The Sound of Many Hands Clapping." *Nature* 403 (6772): 849–850.

Neisser, U. 1967. *Cognitive Psychology*. New York: Appleton-Century-Crofts.

Nestojko, J. F., J. R. Finley, and H. L. Roediger III. 2013. "Extending Cognition to External Agents." *Psychological Inquiry* 24:321–325.

Newell, A. 1976. "Computer Science as Empirical Inquiry: Symbols and Search." *Communications of the ACM* 19:113–126.

Newell, A. 1980. "Physical Symbol Systems." *Cognitive Science* 4 (2): 135–183.

Newell, A. 1990. *Unified Theories of Cognition*. Cambridge, MA: Harvard University Press.

Newell, A., and H. A. Simon. 1972. *Human Problem Solving*. Vol. 104. Englewood Cliffs, NJ: Prentice-Hall.

Niedenthal, P. M. 2007. "Embodying Emotion." *Science* 316 (5827): 1002–1005.

Niedenthal, P. M., L. W. Barsalou, P. Winkielman, S. Krauth-Gruber, and F. Ric. 2005. "Embodiment in Attitudes, Social Perception, and Emotion." *Personality and Social Psychology Review* 9:184–211.

Noë, A. 2004. *Action in Perception*. Cambridge, MA: MIT Press.

Nolen-Hoeksema, S., B. E. Wisco, and S. Lyubomirsky. 2008. "Rethinking Rumination." *Perspectives on Psychological Science* 3 (5): 400–424.

Norman, D. A. 2002. *The Design of Everyday Things*. New York: Basic Books.

Norman, D. A., and D. E. Rumelhart, eds. 1975. *Explorations in Cognition*. San Francisco: Freeman.

Oberman, L. M, and V. S. Ramachandran. 2007. "The Simulating Social Mind: The Role of the Mirror Neuron System and Simulation in the Social and Communicative Deficits of Autism Spectrum Disorders." *Psychological Bulletin* 133:310–327.

Oberman, L. M., P. Winkielman, and V. S. Ramachandran. 2007. "Face to Face: Blocking Facial Mimicry Can Selectively Impair Recognition of Emotional Expressions." *Social Neuroscience* 2 (3–4): 167–178.

Oberman, L. M., P. Winkielman, and V. S. Ramachandran. 2009. "Slow Echo: Facial EMG Evidence for the Delay of Spontaneous, but Not Voluntary, Emotional Mimicry in Children with Autism Spectrum Disorders." *Developmental Science* 12 (4): 510–520.

Onishi, K. H., and R. Baillargeon. 2005. "Do 15-Month-Old Infants Understand False Beliefs?" *Science* 308 (5719): 255–258.

Oosterhof, N. N., A. J. Wiggett, J. Diedrichsen, S. P. Tipper, and P. E. Downing. 2010. "Surface-Based Information Mapping Reveals Crossmodal Vision–Action Representations in Human Parietal and Occipitotemporal Cortex." *Journal of Neurophysiology* 104 (2): 1077–1089.

O'Regan, J. K., and A. Noë. 2001. "A Sensorimotor Account of Vision and Visual Consciousness." *Behavioral and Brain Sciences* 24 (5): 939–973.

Osipov, G. V., J. Kurths, and C. Zhou. 2007. *Synchronization in Oscillatory Networks*. Dordrecht: Springer.

Ostlund, B. D., J. R. Measelle, H. K. Laurent, E. Conradt, and J. C. Ablow. 2017. "Shaping Emotion Regulation: Attunement, Symptomatology, and Stress Recovery within Mother–Infant Dyads." *Developmental Psychobiology* 59 (1): 15–25.

Oullier, O., G. C. De Guzman, K. J. Jantzen, J. Lagarde, and J. A. S. Kelso. 2008. "Social Coordination Dynamics: Measuring Human Bonding." *Social Neuroscience* 3 (2): 178–192.

Paladino, M.-P., M. Mazzurega, F. Pavani, and T. W. Schubert. 2010. "Synchronous Multisensory Stimulation Blurs Self-Other Boundaries." *Psychological Science* 21 (9): 1202–1207.

Parfit, Derek 1984. *Reasons and Persons*. Oxford: Oxford University Press.

Parker, G. 2007. "Defining Melancholia: The Primacy of Psychomotor Disturbance." *Acta Psychiatrica Scandinavica* 115 (supp. 433): 21–30.

Parsaik, A. K., S. S. Mascarenhas, A. Hashmi, L. J. Prokop, V. John, O. Okusaga, and B. Singh. 2016. "Role of Botulinum Toxin in Depression." *Journal of Psychiatric Practice* 22:99–110.

Peckham, A. D., R. K. McHugh, and M. W. Otto. 2010. "A Meta-analysis of the Magnitude of Biased Attention in Depression." *Depression and Anxiety* 27 (12): 1135–1142.

Pellicano, E., M. Maybery, K. Durkin, and A. Maley. 2006. "Multiple Cognitive Capabilities/Deficits in Children with an Autism Spectrum Disorder: 'Weak' Central Coherence and Its Relationship to Theory of Mind and Executive Control." *Development and Psychopathology* 18 (1): 77–98.

Penner-Wilger, M., and M. L. Anderson. 2013. "The Relation between Finger Gnosis and Mathematical Ability: Why Redeployment of Neural Circuits Best Explains the Finding." *Frontiers in Psychology* 4:877.

Peters, U. H. 2000. *Lexikon der Psychiatrie, Psychotherapie, klinischen Psychologie*. München: Urban and Fischer Verlag.

Phillips, J. 2004. "Understanding/Explanation." In *The Philosophy of Psychiatry: A Companion*, edited by J. Radden, 180–190. Oxford: Oxford University Press.

Piccinini, G. 2012. "Computationalism." In *The Oxford Handbook of Philosophy of Cognitive Science*, edited by E. Margolis, R. Samuels, and S. P. Stich, 222–249. Oxford: Oxford University Press.

Piccinini, G., and A. Scarantino. 2010. "Computation vs. Information Processing: Why Their Difference Matters to Cognitive Science." *Studies in History and Philosophy of Science* 41:237–246.

Piek, J. P., and M. J. Dyck. 2004. "Sensory-Motor Deficits in Children with Developmental Coordination Disorder, Attention Deficit Hyperactivity Disorder and Autistic Disorder." *Human Movement Science* 23 (3): 475–488.

Pier, M. P. B., W. Hulstijn, and B. Sabbe. 2004. "Differential Patterns of Psychomotor Functioning in Unmedicated Melancholic and Nonmelancholic Depressed Patients." *Journal of Psychiatric Research* 38 (4): 425–435.

Pikovsky, A. S., M. G. Rosenblum, and J. Kurths. 2001. *Synchronization: A Universal Concept in Nonlinear Science*. Cambridge: Cambridge University Press.

Pinker, S. 2007. *The Stuff of Thought: Language as a Window into Human Nature*. New York: Penguin.

Port, R. F. 2002. "Dynamical Systems Hypothesis in Cognitive Science." In *Encyclopedia of Cognitive Science*, vol. 1, edited by L. Nadel, 1027–1032. London: Nature Publishing Group.

Premack, D., and G. Woodruff. 1978. "Does the Chimpanzee Have a Theory of Mind?" *Behavioral and Brain Sciences* 1 (4): 515–526.

Pulvermüller, F., and L. Fadiga. 2010. "Active Perception: Sensorimotor Circuits as a Cortical Basis for Language." *Nature Reviews Neuroscience* 11:351–360.

Putnam, H. 1975. *Mind, Language, and Reality: Philosophical Papers*. Vol. 2. Cambridge: Cambridge University Press.

Pylyshyn, Z. 1984. *Computation and Cognition*. Cambridge, MA: MIT Press.

Rabelo, A. L. A., V. N. Keller, R. Pilati, and J. M. Wicherts. 2015. "No Effect of Weight on Judgments of Importance in the Moral Domain and Evidence of Publication Bias from a Meta-analysis." *PLOS ONE* 10 (8): e0134808.

Radden, J., and S. Varga. 2013. "Depression Memoirs: Come Meta-analysis." In *The Oxford Handbook of Philosophy and Psychiatry*, edited by K. W. M. Fulford, M. Davies,

R. G. T. Gipps, G. Graham, J. Z. Sadler, G. Stanghellini, and T. Thornton, 99–115. Oxford: Oxford University Press.

Ramsey, W. 1992. "Prototypes and Conceptual Analysis." *Topoi* 11 (1): 59–70.

Ramsey, W. 2014. "Systematicity and Architectural Pluralism." In *The Architecture of Cognition*, edited by P. Calvo and J. Symons, 253–276. Cambridge, MA: MIT Press.

Ramsey, W. 2017. "Must Cognition Be Representational?" *Synthese* 194 (11): 4197–4214.

Ramseyer, F., and W. Tschacher. 2011. "Nonverbal Synchrony in Psychotherapy: Coordinated Body Movement Reflects Relationship Quality and Outcome." *Journal of Consulting and Clinical Psychology* 79 (3): 284.

Ramseyer, F., and W. Tschacher. 2014. "Nonverbal Synchrony of Head- and Body-Movement in Psychotherapy: Different Signals Have Different Associations with Outcome." *Frontiers in Psychology* 5: 979. https://doi.org/10.3389/fpsyg.2014.00979.

Ramseyer, F., and W. Tschacher. 2016. "Movement Coordination in Psychotherapy: Synchrony of Hand Movements is Associated with Session Outcome: A Single-Case Study." *Nonlinear Dynamics, Psychology, and Life Sciences* 20 (2): 145–166.

Rasting, M., and M. E. Beutel. 2005. "Dyadic Affective Interactive Patterns in the Intake Interview as a Predictor of Outcome." *Psychotherapy Research* 15 (3): 188–198.

Ratcliffe, M. 2008. *Feelings of Being: Phenomenology, Psychiatry and the Sense of Reality*. Oxford: Oxford University Press.

Ratcliffe, M. 2010. "The Phenomenology of Mood and the Meaning of Life." In *Oxford Handbook on Emotion*, edited by P. Goldie, 349–371. Oxford: Oxford University Press.

Ray, E., and C. Heyes. 2011. "Imitation in Infancy: The Wealth of the Stimulus." *Developmental Science* 14 (1): 92–105.

Reck, C., A. Hunt, R. Weiss, T. Fuchs, E. Möhler, G. Downing, E. Z. Tronick, and C. H. Mundt. 2004. "Interactive Regulation of Affect in Postpartum Depressed Mothers and Their Infants." *Psychopathology* 3:272–280.

Reddy, V. 2008. *How Infants Know Minds*. Cambridge, MA: Harvard University Press.

Reuven, O., N. Liberman, and R. Dar. 2014. "The Effect of Physical Cleaning on Threatened Morality in Individuals with Obsessive-Compulsive Disorder." *Clinical Psychological Science* 2 (2): 224–229.

Richardson, D. C., and R. Dale. 2005. "Looking to Understand: The Coupling between Speakers' and Listeners' Eye Movements and Its Relationship to Discourse Comprehension." *Cognitive Science* 29 (6): 1045–1060.

Richardson, M. J., K. Shockley, B. R. Fajen, M A. Riley, and M. T. Turvey. 2008. "Ecological Psychology: Six Principles for an Embodied-Embedded Approach to Behavior." In *Handbook of Cognitive Science: An Embodied Approach*, edited by P. Calvo and T. Gomila, 159–187. Amsterdam: Elsevier.

Richardson, M. J., K. Shockley, M. R. Riley, B. R. Fajen, and M. T. Turvey. 2008. "Ecological Psychology: Six Principles for an Embodied-Embedded Approach to Behavior." In *Elsevier Handbook of New Directions in Cognitive Science Section I. The Embodied Architecture of Cognition: Conceptual Issues*, edited by P. Calvo and T. Gomila, 161–187. San Diego: Elsevier.

Rimé, B. 2009. "Emotion Elicits the Social Sharing of Emotion: Theory and Empirical Review." *Emotion Review* 1 (1): 60–85.

Rinehart, N. J., J. L. Bradshaw, A. V. Brereton, and B. J. Tonge. 2001. "Movement Preparation in High-Functioning Autism and Asperger Disorder: A Serial Choice Reaction Time Task Involving Motor Reprogramming." *Journal of Autism and Developmental Disorders* 31 (1): 79–88.

Riskind, J. H. 1983. "Nonverbal Expressions and the Accessibility of Life Experience Memories: A Congruence Hypothesis." *Social Cognition* 2 (1): 62–86.

Risko, E. F., and S. J. Gilbert. 2016. "Cognitive Offloading." *Trends in Cognitive Sciences* 20 (9): 676–688.

Risko, E. F., Medimorec, S., Chisholm, J. D., and Kingstone, A. 2014. "Rotating with Rotated Text: A Natural Behavior Approach to Investigating Cognitive Offloading." *Cognitive Science* 38:537–564.

Ritchie, L. D. 2013. *Metaphor*. Cambridge: Cambridge University Press.

Rizzolatti, G., and M. A. Arbib. 1998. "Language within Our Grasp." *Trends in Neurosciences* 21 (5): 188–194.

Rizzolatti, G., and C. Sinigaglia. 2010. "The Functional Role of the Parieto-Frontal Mirror Circuit: Interpretations and Misinterpretations." *Nature Reviews Neuroscience* 11 (4): 264–274.

Robbins, P., and M. Aydede. 2009. "A Short Primer on Situated Cognition." *The Cambridge Handbook of Situated Cognition*, edited by P. Robbins and M. Aydede, 3–10. Cambridge: Cambridge University Press.

Rogers, S. J., and B. F. Pennington. 1991. "A Theoretical Approach to the Deficits in Infantile Autism." *Development and Psychopathology* 3 (2): 137–162.

Rogers, T. T., and J. L. McClelland. 2014. "Parallel Distributed Processing at 25: Further Explorations in the Microstructure of Cognition." *Cognitive Science* 38 (6): 1024–1077.

Röhricht, F. 2009. "Body Oriented Psychotherapy: The State of the Art in Empirical Research and Evidence-Based Practice: A Clinical Perspective." *Body, Movement and Dance in Psychotherapy* 4 (2): 135–156.

Romero, F. 2015. "Why There Isn't Inter-level Causation in Mechanisms." *Synthese* 192 (11): 3731–3755.

Rosch, E. 1973. "Natural Categories." *Cognitive Psychology* 4:328–350.

Rosch, E., and C. B. Mervis. 1975. Family Resemblances: Studies in the Internal Structure of Categories.*Cognitive Psychology* 7 (4): 573-605.

Rowlands, M. 1999. *The Body in Mind: Understanding Cognitive Processes*. Cambridge: Cambridge University Press.

Rowlands, M. 2006. *Body Language: Representation in Action*. Cambridge, MA: MIT Press.

Rowlands, M. 2008. "Extended Cognition and the Mark of the Cognitive." *Philosophical Psychology* 22 (1): 1–19.

Rowlands, M. 2010. *The New Science of the Mind: From Extended Mind to Embodied Phenomenology*. Cambridge, MA: MIT Press.

Rowlands, M. 2015. "Bringing Philosophy Back: 4e Cognition and the Argument from Phenomenology." In *Philosophy of Mind and Phenomenology: Conceptual and Empirical Approaches*, edited by D. O. Dahlstrom, A. Elpidorou, and W. Hopp, 310–325. London: Routledge.

Rumelhart, D. E. 1989. "The Architecture of Mind: A Connectionist Approach." In *Foundations of Cognitive Science*, edited by M. Posner, 133–159. Cambridge, MA: MIT Press.

Rumelhart, D. E., J. L. McClelland, and P. R. Group. 1986. *Foundations*. Vol. 1 of *Parallel Distributed Processing: Explorations in the Microstructures of Cognition*. Cambridge, MA: MIT Press.

Rümke, H. C. 1941. "The Nuclear Symptom of Schizophrenia and the Praecoxfeeling." Translated by J. Neeleman. *History of Psychiatry* 1:331–341.

Rupert, R. D. 2004. "Challenges to the Hypothesis of Extended Cognition." *Journal of Philosophy* 101 (8): 389–428.

Rupert, R. D. 2009. *Cognitive Systems and the Extended Mind*. Oxford: Oxford University Press.

Rupert, R. D. 2010. "Extended Cognition and the Priority of Cognitive Systems." *Cognitive Systems Research* 11 (4): 343–356.

Rupert, Robert D. 2013. "Memory, Natural Kinds, and Cognitive Extension; or, Martians Don't Remember, and Cognitive Science Is Not about Cognition." *Review of Philosophy and Psychology* 4 (1): 25-47.

Sameroff, A., ed. 2009. *The Transactional Model of Development: How Children and Contexts Dhape Each Other*. Washington, DC: American Psychological Association.

Samuels, R., E. Margolis, and S. P. Stich. 2012. "Introduction: Philosophy and Cognitive Science." In *The Oxford Handbook of Philosophy of Cognitive Science*, edited by E. Margolis, R. Samuels, and S. P. Stich, 3–18. Oxford: Oxford University Press.

Santos, L. R., A. G. Nissen, and J. A. Ferrugia. 2006. "Rhesus Monkeys, Macaca Mulatta, Know What Others Can and Cannot Hear." *Animal Behavior* 71:1175–1181.

Sass, L., and J. Parnas. 2017. "Thought Disorder, Subjectivity, and the Self." *Schizophrenia Bulletin* 43:497–502.

Saxbe, D., and R. L. Repetti. 2010. "For Better or Worse? Coregulation of Couples' Cortisol Levels and Mood States." *Journal of Personality and Social Psychology* 98 (1): 92.

Sbarra, D. A., and C. Hazan. 2008. "Coregulation, Dysregulation, Self-regulation: An Integrative Analysis and Empirical Agenda for Understanding Adult Attachment, Separation, Loss and Recovery." *Personality and Social Psychology Review* 12:141–167.

Scheeren, A. M., M. de Rosnay, H. M. Koot, and S. Begeer. 2013. "Rethinking Theory of Mind in High-Functioning Autism Spectrum Disorder." *Journal of Child Psychology and Psychiatry* 54 (6): 628–635.

Schmidt, R. C., and B. O'Brien. 1997. "Evaluating the Dynamics of Unintended Interpersonal Coordination." *Ecological Psychology* 9 (3): 189–206.

Schmidt, R. C., and M. Richardson. 2008. "Dynamics of Interpersonal Coordination." In *Coordination: Neural, Behavioral and Social Dynamics: Understanding Complex Systems*, edited by A. Fuchs and V. K. Jirsa, 281–308. Berlin: Springer.

Schnall, S., J. Benton, and S. Harvey. 2008. "With a Clean Conscience: Cleanliness Reduces the Severity of Moral Judgments." *Psychological Science* 19:1219–1222.

Schneider, D., V. P. Slaughter, A. P. Bayliss, and P. E. Dux. 2013. "A Temporally Sustained Implicit Theory of Mind Deficit in Autism Spectrum Disorders." *Cognition* 129 (2): 410–417.

Schneider, I., B. Rutjens, N. Jostmann, and D. Lakens. 2011. "Weighty Matters: Importance Literally Feels Heavy." *Social Psychological and Personality Science* 2: 474–478.

Schneider, W. 1987. "Connectionism: Is It a Paradigm Shift for Psychology?" *Behavior Research Methods, Instruments, and Computers* 19:73–83.

Schnitzspahn, K. M., A. Ihle, J. D. Henry, P. G. Rendell, and M. Kliegel. 2011. "The Age-Prospective Memory-Paradox: An Exploration of Possible Mechanisms." *International Psychogeriatrics* 23 (4): 583–592.

Scholl, B. J., and A. M. Leslie. 1999. "Modularity, Development and 'Theory of Mind.'" *Mind & Language* 14 (1): 131–153.

Schrijvers, D., W. Hulstijn, and B. G. Sabbe. 2008. "Psychomotor Symptoms in Depression: A Diagnostic, Pathophysiological and Therapeutic Tool." *Journal of Affective Disorders* 109 (1): 1–20.

Schroer, Jeanine, and Robert Schroer. 2013. "Two Potential Problems with Philosophical Intuitions: Muddled Intuitions and Biased Intuitions." *Philosophia* 41 (4): 1263–1281.

Schubert, T. W. 2005. "Your Highness: Vertical Positions as Perceptual Symbols of Power." *Journal of Personality and Social Psychology* 89 (1): 1–21.

Schukraft, Jason. 2016. "Carving Intuition at Its Joints." *Metaphilosophy* 47, no. 3: 326–352.

Searle, J. R. 1992. *The Rediscovery of the Mind.* Cambridge, MA: MIT Press.

Sellars, W. 1956. "Empiricism and the Philosophy of Mind." In *Minnesota Studies in the Philosophy of Science,* vol. 1, ed. Herbert Feigl and Michael Scriven, 253–329. Minneapolis: University of Minnesota Press.

Semin, G. R., and J. T. Cacioppo. 2008. "Grounding Social Cognition: Synchronization, Entrainment, and Coordination." In *Embodied Grounding: Social, Cognitive, Affective,* and *Neuroscientific Approaches,* edited by G. R. Semin and E. R. Smith, 119–147. New York: Cambridge University Press.

Senju, A. 2012. "Spontaneous Theory of Mind and Its Absence in Autism Spectrum Disorders." *Neuroscientist* 18 (2): 108–113.

Senju, A., V. Southgate, Y. Miura, T. Matsui, T. Hasegawa, Y. Tojo, et al. 2010. "Absence of Spontaneous Action Anticipation by False Belief Attribution in Children with Autism Spectrum Disorder." *Development and Psychopathology* 22 (2): 353–360.

Senju, A., V. Southgate, S. White, and U. Frith. 2009. "Mindblind Eyes: An Absence of Spontaneous Theory of Mind in Asperger Syndrome." *Science* 325 (5942): 883–885.

Shafir, T. 2015. "Movement-Based Strategies for Emotion Regulation." In *Handbook on Emotion Regulation: Processes, Cognitive Effects and Social Consequences,* edited by M. L. Bryant, 231–249. New York: Nova Science Publishers.

Shapiro, L. A. 2011. *Embodied Cognition.* New York: Routledge.

Shapiro, L. A. 2014. "Introduction." In *The Routledge Handbook of Embodied Cognition,* edited by L. A. Shapiro, 1–8. New York: Routledge.

Sharrock, W., and J. Coulter. "ToM: A Critical Commentary." *Theory & Psychology* 14(5): 579–600.

Shi, Z., A. W. Taylor, G. Wittert, R. Goldney, and T. K. Gill. 2010. "Soft Drink Consumption and Mental Health Problems among Adults in Australia." *Public Health Nutrition* 13 (7): 1073–1079.

Shockley, K., M.-V. Santana, and C. A. Fowler. 2003. "Mutual Interpersonal Postural Constraints Are Involved in Cooperative Conversation." *Journal of Experimental Psychology: Human Perception and Performance* 29 (2): 326–332.

Singer, T., B. Seymour, J. O'doherty, H. Kaube, R. J. Dolan, and C. D. Frith. 2004. "Empathy for Pain Involves the Affective but Not Sensory Components of Pain." *Science* 303 (5661): 1157–1162.

Slater, M. H. 2005. "Monism on the One Hand, Pluralism on the Other." *Philosophy of Science* 72 (1): 22–42.

Smith, A. 2009. "The Empathy Imbalance Hypothesis of Autism: A Theoretical Approach to Cognitive and Emotional Empathy in Autistic Development." *Psychological Record* 59:489–510.

Smolensky, P. 1994. *Computational Models of Mind: A Companion to Philosophy of Mind*. Oxford: Blackwell.

Smolensky, P. 1995. "Reply: Constituent Structure and Explanation in an Integrated Connectionist/Symbolic Architecture." In *Connectionism: Debates on Psychological Explanation*, edited by C. Macdonald and G. Macdonald, 223–290. Oxford: Oxford University Press.

Smolensky, P., and G. Legendre. 2006. *Cognitive Architecture*. Vol. 1, *The Harmonic Mind: From Neural Computation to Optimality-Theoretic Grammar*. Cambridge, MA: MIT Press.

Sobin, C., L. Mayer, and J. Endicott. 1998. "The Motor Agitation and Retardation Scale: A Scale for the Assessment of Motor Abnormalities in Depressed Patients." *The Journal of Neuropsychiatry and Clinical Neurosciences* 10 (1): 85–92.

Song, H., and R. Baillargeon. 2008. "Infants' Reasoning about Others' False Perceptions." *Developmental Psychology* 44 (6): 1789.

Sosa, E. 2007. "Experimental Philosophy and Philosophical Intuition." *Philosophical Studies* 132 (1): 99–107.

Southgate, V., C. Chevallier, and G. Csibra. 2010. "Seventeen-Month-Olds Appeal to False Beliefs to Interpret Others' Referential Communication." *Developmental Science* 13 (6): 907–912.

Southgate, V., A. Senju, and G. Csibra. 2007. "Action Anticipation through Attribution of False Belief by 2-Year-Olds." *Psychological Science* 18 (7): 587–592.

Sparrow, B., J. Liu, and D. M. Wegner. 2011. "Google Effects on Memory: Cognitive Consequences of Having Information at Our Fingertips." *Science* 333 (6043): 776–778.

Spivey, M. 2007. *The Continuity of Mind.* Oxford: Oxford University Press.

Sprevak, M. 2009. "Extended Cognition and Functionalism." *Journal of Philosophy* 106 (9): 503–527.

Sprevak, M. 2010. Computation and cognitive science. *Studies in History and Philosophy of Science Part A* 41 (3): 223–226.

Sprevak, M. 2011. "Neural Sufficiency, Reductionism, and Cognitive Neuropsychiatry." *Philosophy, Psychiatry, and Psychology* 18 (4): 339–344.

Srinivasan, M., and S. Carey. 2010. "The Long and the Short of It: On the Nature and Origin of Functional Overlap between Representations of Space and Time." *Cognition* 116 (2): 217–241.

Staples, K. L., and G. Reid. 2010. "Fundamental Movement Skills and Autism Spectrum Disorders." *Journal of Autism and Developmental Disorders* 40 (2): 209–217.

Stein, D. J. 1992. "Cognitive Science and Psychiatry: An Overview." *Intergrative Psychiatry* 8:109–116.

Sterelny, K. 2003. *Thought in a Hostile World: The Evolution of Human Cognition.* Oxford: Blackwell Publishing.

Sterelny, K. 2010. "Minds: Extended or Scaffolded?" *Phenomenology and the Cognitive Sciences* 9 (4): 465–481.

Sterelny, K. 2012. *The Evolved Apprentice.* Cambridge, MA: MIT Press.

Sterelny, K., and P. Griffiths. 2010. "What Is Life?" In *The Nature of Life: Classical and Contemporary Perspectives from Philosophy and Science,* edited by M. A. Bedau and C. E. Cleland, 355–359. Cambridge: Cambridge University Press.

Stern, D. 1985. *The Interpersonal World of the Infant.* New York: Basic Books.

Stich, S., and W. Buckwalter. 2011. "Gender and the Philosophy Club." *The Philosophers' Magazine* 52 (52): 60–65.

Storm, B. C., and S. M. Stone. 2015. "Saving-Enhanced Memory: The Benefits of Saving on the Learning and Remembering of New Information." *Psychological Science* 26 (2): 182–188.

Strawson, P. 1992. *Analysis and Metaphysics: An Introduction to Philosophy.* Oxford: Oxford University Press.

Surian, L., S. Caldi, and D. Sperber. 2007. "Attribution of Beliefs by 13-Month-Old Infants." *Psychological Science* 18 (7): 580–586.

Sutera, S., J. Pandey, E. L. Esser, M. Rosenthal, L. B. Wilson, M. Barton, and D. Fein. 2007. "Predictors of Optimal Outcome in Toddlers Diagnosed with Autism Spectrum Disorders." *Journal of Autism and Developmental Disorders* 37 (1): 98–107.

Sutton, J. 2010. "Exograms and Interdisciplinarity: History, the Extended Mind, and the Civilizing Process." In *The Extended Mind*, edited by R. Menary, 189–226. Cambridge, MA: MIT Press.

Sutton, J. 2015. "Scaffolding Memory: Themes, Taxonomies, Puzzles." In *Contextualizing Human Memory: An Interdisciplinary Approach to Understanding How Individuals and Groups Remember the Past*, edited by L. Bietti and C. Stone, 187–205. New York: Routledge.

Sutton, J., C. B. Harris, P. G. Keil, and A. J. Barnier. 2010. "The Psychology of Memory, Extended Cognition, and Socially Distributed Remembering." *Phenomenology and the Cognitive Sciences* 9 (4): 521–560.

Swain, S., J. Alexander, and J. M. Weinberg. 2008."The Instability of Philosophical Intuitions: Running Hot and Cold on Truetemp." *Philosophy and Phenomenological Research* 76 (1): 138–155.

Szabo Gendler, T. 2007. "Philosophical Thought Experiments, Intuitions, and Cognitive Equilibrium." *Midwest Studies in Philosophy* 31:68–89.

Szymkow, A., J. Chandler, S. T. IJzerman, M. Parzuchowski, and B. Wojciszke. 2013. "Warmer Hearts, Warmer Rooms." *Social Psychology* 44 (2): 167–176.

Teasdale, J. D., and P. J. Barnard. 1993. "Affect, Cognition and Change: Re-modelling Depressive Thought." Hove, UK: Lawrence Erlbaum Associates.

Teitelbaum, P., O. Teitelbaum, J. Nye, J. Fryman, and R. G. Maurer. 1998. "Movement Analysis in Infancy May Be Useful for Early Diagnosis of Autism." *Proceedings of the National Academy of Sciences* 95 (23): 13982–13987.

Thagard, P. 2005. *Mind: Introduction to Cognitive Science*. Vol. 4. Cambridge, MA: MIT Press.

Thagard, P. 2009. "Why Cognitive Science Needs Philosophy and Vice Versa." *Topics in Cognitive Science* 1 (2): 237–254.

Theiner, G., C. Allen, and R. L. Goldstone. 2010. "Recognizing Group Cognition." *Cognitive Systems Research* 11 (4): 378–395.

Thelen, E., and L. Smith. 1994. *A Dynamic Systems Approach to the Development of Cognition and Action*. Cambridge, MA: MIT Press.

Thomasson, A. 2014. "The Easy Approach to Ontology: A Defense." In *Philosophical Methodology: The Armchair or the Laboratory*, edited by M. C. Haug, 107–125. London: Routledge.

Thompson, E. 2007. *Mind in Life: Biology, Phenomenology, and the Sciences of Mind.* Cambridge, MA: Harvard University Press.

Thompson, E., and F. J. Varela. 2001. "Radical Embodiment: Neural Dynamics and Consciousness." *Trends in Cognitive Sciences* 5 (10): 418–425.

Thompson, R. A. 1994. "Emotion Regulation: A Theme in Search of Definition." *Monographs of the Society for Research in Child Development* 59 (2-3): 25–52.

Tienson, J. L. 1988. "An Introduction to Connectionism." *Southern Journal of Philosophy* 26:1–16.

Tollefsen, D. P. 2006. "From Extended Mind to Collective Mind." *Cognitive Systems Research* 7:140–150.

Tomasello, M. 1999. "Social Cognition before the Revolution." In *Early Social Cognition: Understanding Others in the First Months of Life*, edited by P. Rochat, 301–314. Mahwah, NJ: Lawrence Erlbaum Associates.

Tooby, J., and L. Cosmides. 2008. "The Evolutionary Psychology of the Emotions and Their Relationship to Internal Regulatory Variables." In *Handbook of Emotions*, 3rd ed., edited by M. Lewis, J. M. Haviland-Jones, and L. F. Barrett, 114–137. New York: Guilford Press.

Torres, E. B., M. Brincker, R. W. Isenhower, P. Yanovich, K. A. Stigler, J. I. Nurnberger, D. N. Metaxas, et al. 2013. "Autism: The Micro-movement Perspective." *Frontiers in Integrative Neuroscience* 7:32.

Trevarthen, C. 2002. "Origins of Musical Identity: Evidence from Infancy for Musical Social Awareness." In *Musical Identities*, edited by R. A. R. MacDonald, D. J. Hargreaves, and D. Miell, 21–38. Oxford: Oxford University Press.

Trevarthen, C., and S. Daniel. 2005. "Disorganized Rhythm and Synchrony: Early Signs of Autism and Rett Syndrome." *Brain and Development* 27:S25–S34.

Tronick, E. Z. 1989. "Emotions and Emotional Communication in Infants." *American Psychologist* 44:112–119.

Tronick, E. Z. 1998. "Dyadically Expanded States of Consciousness and the Process of Therapeutic Change." *Infant Mental Health Journal* 19 (3): 290–299.

Tronick, E. Z. 2002. "A Model of Infant Mood States: Long Lasting Organizing Affective States and Emotional Representational Processes without Language or Symbols." *Psychoanalytic Dialogues* 12 (1): 73–99

Tronick, E. Z., H. Als, and L. Adamson. 1979. "The Communicative Structure of Face-to-Face Interaction." In *Before Speech: The Beginnings of Human Communication*, edited by M. Bullowa, 349–372. Cambridge: Cambridge University Press.

Tronick, E. Z., and M. Beeghly. 2011. "Infants' Meaning-Making and the Development of Mental Health Problems." *American Psychologist* 66:107–119.

Tronick, E. Z., and A. F. Gianino. 1986. "The Transmission of Maternal Disturbance to the Infant." *New Directions for Child and Adolescent Development* 34:5–11.

Tryon, W. W. 2014. *Cognitive Neuroscience and Psychotherapy: Network Principles for a Unified Theory*. London: Academic Press.

Tschacher, W., and M. Pfammatter. 2016. "Embodiment in Psychotherapy: A Necessary Complement to the Canon of Common Factors." *European Psychotherapy* 13:9–25.

Tschacher, W., G. M. Rees, and F. Ramseyer. 2014. "Nonverbal Synchrony and Affect in Dyadic Interactions." *Frontiers in Psychology* 5:1323.

Tversky, B. 2011. "Visualizing Thought." *Topics in Cognitive Science* 3 (3): 499–535.

Uithol, S., and V. Gallese. 2015. "The Role of the Body in Social Cognition." *WIREs: Cognitive Science* 6 (5): 453–460.

Ullrich, H., L. Kranaster, E. Sigges, J. Andrich, and A. Sartorius. 2012. "Ultra-High-Frequency Left Prefrontal Transcranial Magnetic Stimulation as Augmentation in Severely Ill Patients with Depression: A Naturalistic Sham-Controlled, Double-Blind, Randomized Trial." *Neuropsychobiology* 66:141–148.

Umilta, L., E. Kohler, V. Gallese, L. Fogassi, L. Fadiga, C. Keysers, and G. Rizzolatti. 2001. "I Know What You Are Doing: A Neurophysiological Study." *Neuron* 31:155–165.

Ungvari, G. S., H. F. K. Chiu, H. C. M. Leung, L. Y. Chow, Y. Hong, E. M. P. So, and F. C. K. Lum. 1997. "Conceptual Issues in the Diagnosis of Schizophrenic Psychoses." *Hong Kong Journal of Psychiatry* 7 (2): 3–8.

van Eck, D. 2018. "Constitutive Relevance in Cognitive Science: The Case of Eye Movements and Cognitive Mechanisms." *Studies in History and Philosophy of Science Part A*. https://doi.org/10.1016/j.shpsa.2018.05.013.

van Eck, D., and H. L. de Jong. 2016. "Mechanistic Explanation, Cognitive Systems Demarcation, and Extended Cognition." *Studies in History and Philosophy of Science Part A* 59:11–21.

van Gelder, T. J. 1995. "What Might Cognition Be, If Not Computation?" *Journal of Philosophy* 91:345–381.

van Gelder, T. J. 1998a. "The Dynamical Hypothesis in Cognitive Science." *Behavioral and Brain Sciences* 21 (5): 615–628.

van Gelder, T. J. 1998b. "The Roles of Philosophy in Cognitive Science." *Philosophical Psychology* 11 (2): 117–136.

van Gelder, T. J., and R. Port. 1995. "It's about Time: A Perspective to Dynamical System Approach to Cognition." In *Mind as Motion*, edited by R. Port and T. van Gelder, 1–43. Cambridge, MA: MIT Press.

van Holland, M. 2013. "Extended Cognition and the Mark of the Cognitive: Prospects of a Proper Function-Based Approach." Eindhoven: Technische Universiteit Eindhoven. https://doi.org/10.6100/IR759659.

van Rooij, I., R. M. Bongers, and W. F. G. Haselager. 2002. "A Non-representational Approach to Imagined Action." *Cognitive Science* 26 (3): 345–375.

van Ulzen, N. R., C. J. Lamoth, A. Daffertshofer, G. R. Semin, and P. J. Beek. 2008. "Characteristics of Instructed and Uninstructed Interpersonal Coordination while Walking Side-by-Side." *Neuroscience Letters* 432 (2): 88–93.

Varela, F. J., E. Thompson, and E. Rosch. 1991. *The Embodied Mind: Cognitive Science and Human Experience*. Cambridge, MA: MIT Press.

Varga, S. 2013. "Vulnerability to Psychosis, I-Thou Intersubjectivity and the Praecox-Feeling." *Phenomenology and the Cognitive Sciences* 12 (1): 131–143.

Varga, S. 2014. "Cognition, Representations and Embodied Emotions: Investigating Cognitive Theory." *Erkenntnis* 79 (1): 165–190.

Varga, S. 2015. *Naturalism, Interpretation, and Mental Disorder*. Oxford: Oxford University Press.

Varga, S. 2016. "Interaction and Extended Cognition." *Synthese* 193 (8): 2469–2496.

Varga, S. 2017. "Demarcating the Realm of Cognition." *Journal for General Philosophy of Science* 49 (3): 435–450.

Varga, S. 2018a. "Embodied Concepts and Disordered Minds." *Journal of Medicine and Philosophy* 43 (2): 241–260.

Varga, S. 2018b. "Embodiment and Moral Reasoning." In *The Oxford Handbook of 4E Cognition*, edited by L. De Bruin, A. Newen, and S. Gallagher, 863–874. Oxford: Oxford University Press.

Varlet, M., L. Marin, J. Lagarde, and B. G. Bardy. 2011. "Social Postural Coordination." *Journal of Experimental Psychology: Human Perception and Performance* 37 (2): 473–483.

Veenstra, L., I. K. Schneider, and S. L. Koole. 2017. "Embodied Mood Regulation: The Impact of Body Posture on Mood Recovery, Negative Thoughts, and Mood-Congruent Recall." *Cognition and Emotion* 31 (7): 1361–1376.

Vilensky, J. A., A. R. Damasio, and R. G. Maurer. 1981. "Gait Disturbances in Patients with Autistic Behavior: A Preliminary Study." *Archives of Neurology* 38 (10): 646–649.

Vivanti, G., and S. J. Rogers. 2011. "Action Understanding and Social Learning in Autism: A Developmental Perspective." *Life Span and Disability* 14 (1): 7–29.

Walter, S., and L. Kästner. 2012. The Where and What of Cognition: The Untenability of Cognitive Agnosticism. *Cognitive Systems Research*, 13 (1): 12–23.

Walther, S., K. Stegmayer, J. Sulzbacher, T. Vanbellingen, R. Müri, W. Strik, and S. Bohlhalter. 2015. "Nonverbal Social Communication and Gesture Control in Schizophrenia." *Schizophrenia Bulletin* 41(2): 338–345.

Wang, Y. G., Y. Q. Wang, S. L. Chen, C. Y. Zhu, and K. Wang. 2008. "Theory of Mind Disability in Major Depression with or without Psychotic Symptoms: A Componential View." *Psychiatry Research* 161:153–161.

Ward, D. 2016. "Achieving Transparency: An Argument for Enactivism." *Philosophy and Phenomenological Research* 93 (3): 650–680.

Warneken, F., and M. Tomasello. 2006. "Altruistic Helping in Human Infants and Young Chimpanzees." *Science* 311 (5765): 1301–1303.

Way, B. M., S. E. Taylor, and N. I. Eisenberger. 2009. "Variation in the μ-opioid Receptor Gene (OPRM1) Is Associated with Dispositional and Neural Sensitivity to Social Rejection." *Proceedings of the National Academy of Sciences* 106 (35): 15079–15084.

Wegner, D. M. 1995. "A Computer Network Model of Human Transactive Memory." *Social Cognition* 13 (3): 319–339.

Wegner, D. M., R. Erber, and P. Raymond. 1991. "Transactive Memory in Close Relationships." *Journal of Personality and Social Psychology* 61 (6): 923–929.

Weinberg, J. 2017. "What Is Negative Experimental Philosophy Good For?" In *The Cambridge Companion to Philosophical Methodology*, edited by G. D'Oro and S. Overgaard, 141–160. Cambridge: Cambridge University Press.

Weinberg, J. S., S. Nichols, and S. Stich. 2001. "Normativity and Epistemic Intuitions." *Philosophical Topics* 29 (1–2): 429–460.

Weinberg, M. K., K. L. Olson, M. Beeghly, and E. Z. Tronick. 2006. "Making Up Is Hard to Do, Especially for Mothers with High Levels of Depressive Symptoms and Their Infant Sons." *Journal of Child Psychology and Psychiatry* 47:670–683.

Weinberg, M. K., and E. Z. Tronick. 1994. "Beyond the Face: An Empirical Study of Infant Affective Configurations of Facial, Vocal, Gestural, and Regulatory Behaviors." *Child Development* 65:1495–1507.

Wheatley, T., and J. Haidt. 2005. "Hypnotically Induced Disgust Makes Moral Judgments More Severe." *Psychological Science* 16:780–784.

Wheatley, T., O. Kang, C. Parkinson, and C. E. Looser. 2012. "From Mind Perception to Mental Connection: Synchrony as a Mechanism for Social Understanding." *Social and Personality Psychology Compass* 6 (8): 589–606.

Wheeler, M. 2005. *Reconstructing the Cognitive World: The Next Step*. Cambridge, MA: MIT Press.

Wheeler, M. 2010. "In Defense of Extended Functionalism." In *The Extended Mind*, edited by R. Menary, 245–270. Cambridge, MA: MIT Press.

Whiten, A. 2000. "Primate Culture and Social Learning." *Cognitive Science* 24 (3): 477–508.

Whyatt, C., and C. M. Craig. 2013a. "Interceptive Skills in Children Aged 9–11 Years, Diagnosed with Autism Spectrum Disorder." *Research in Autism Spectrum Disorders* 7 (5): 613–623.

Whyatt, C., and C. M. Craig. 2013b. "Sensory-Motor Problems in Autism." *Frontiers in Integrative Neuroscience* 7:51.

Whyte, J., and A. Owens. 1989. "Language and Symbolic Object Play: Some Findings from a Study of Autistic Children." *Irish Journal of Psychology* 10:317–332.

Wicker, B., C. Keysers, J. Plailly, J.-P. Royet, V. Gallese, and G. Rizzolatti. 2003. "Both of Us Disgusted in My Insula: The Common Neural Basis of Seeing and Feeling Disgust." *Neuron* 40 (3): 655–664.

Wilkes, C., R. Kydd, M. Sagar, and E. Broadbent. 2017. "Upright Posture Improves Affect and Fatigue in People with Depressive Symptoms." *Journal of Behavior Therapy and Experimental Psychiatry* 54:143–149.

Wilkes, K. V. 1988. *Real People: Personal Identity without Thought Experiments*. Oxford: Oxford University Press.

Wilkowski, B. M., B. P. Meier, M. D. Robinson, M. S. Carter, and R. Feltman. 2009. "Hot-Headed Is More than an Expression: The Embodied Representation of Anger in Terms of Heat." *Emotion* 9:464–477.

Williams, J. H. 2008. "Self–Other Relations in Social Development and Autism: Multiple Roles for Mirror Neurons and Other Brain Bases." *Autism Research* 1 (2): 73–90.

Williams, J. H. G., G. D. Waiter, A. Gilchrist, D. I. Perrett, A. D. Murray, and A. Whiten. 2006. "Neural Mechanisms of Imitation and Mirror Neuron Functioning in Autistic Spectrum Disorder." *Neuropsychologia* 44:610–621.

Williams, J. H. G., A. Whiten, T. Suddendorf, and D. I. Perrett. 2001. "Imitation, Mirror Neurons and Autism." *Neuroscience and Biobehavioral Reviews* 25 (4): 287–295.

Williams, J. M. G., F. N. Watts, C. MacLeod, and A. Mathews. 1988. *Cognitive Psychology and Emotional Disorders*. New York: Wiley.

Williams, J. M., F. N. Watts, C. MacLeod, and A. Mathews. 1997. *Cognitive Psychology and Emotional Disorders*. 2nd ed. Chichester: Wiley.

Williams, L. E., and J. A. Bargh. 2008a. "Experiencing Physical Warmth Promotes Interpersonal Warmth." *Science* 322 (5901): 606–607.

Williams, L. E., and J. A. Bargh. 2008b. "Keeping One's Distance: The Influence of Spandatial Distance Cues on Affect and Evaluation." *Psychological Science* 19 (3): 302–308.

Williams, L. E., J. Y. Huang, and J. A. Bargh. 2009. "The Scaffolded Mind: Higher Mental Processes Are Grounded in Early Experience of the Physical World." *European Journal of Social Psychology* 39 (7): 1257–1267.

Williamson, T. 2004. "Philosophical 'Intuitions' and Scepticism about Judgement." *Dialectica* 58 (1): 109–153.

Williamson, T. 2007. *The Philosophy of Philosophy*. Oxford: Blackwell.

Wilson, A. D., and S. Golonka. 2013. "Embodied Cognition Is Not What You Think It Is." *Frontiers in Psychology* 4:58.

Wilson, J. 2016. "VII-Internal and External Validity in Thought Experiments." *Proceedings of the Aristotelian Society* 116 (2): 127–152.

Wilson, M. 2002. "Six Views of Embodied Cognition." *Psychonomic Bulletin & Review* 9:625–636.

Wilson, R. A. 1994. "Wide Computationalism." *Mind* 103 (411): 351–372.

Wilson, R. A. 2004. *Boundaries of the Mind: The Individual in the Fragile Sciences: Cognition*. Cambridge: Cambridge University Press.

Wilson, R. A. 2005. "Collective Memory, Group Minds, and the Extended Mind Thesis." *Cognitive Processing* 6: 227–236.

Wiltermuth, S. S. 2012. "Synchronous Activity Boosts Compliance with Requests to Aggress." *Journal of Experimental Social Psychology* 48 (1): 453–456.

Wimmer, H., and J. Perner. 1983. "Beliefs about Beliefs: Representation and Constraining Function of Wrong Beliefs in Young Children's Understanding of Deception." *Cognition* 13 (1): 103–128.

Wimsatt, W. C., and J. R. Griesemer. 2007. "Reproducing Entrenchments to Scaffold Culture: The Central Role of Development in Cultural Evolution." In *Integrating Evolution and Development: From Theory to Practice*, edited by R. Sansome and R. Brandon, 228–323. Cambridge, MA: MIT Press.

Winer, E. S., and T. Salem. 2016. "Reward Devaluation: Dot-Probe Meta-analytic Evidence of Avoidance of Positive Information in Depressed Persons." *Psychological Bulletin* 142 (1): 18–78.

Wolf, F., M. Brüne, and H.-J. Assion. 2010. "Theory of Mind and Neurocognitive Functioning in Patients with Bipolar Disorder." *Bipolar Disorders* 12:657–666.

Wollmer, M. A. 2014. "Treatment of Major Depressive Disorder Using Botulinum Toxin A: A 24-Week Randomized, Double-Blind, Placebo-Controlled Study." *Journal of Clinical Psychiatry* 75 (8): 837–844.

Wollmer, M. A., C. de Boer, N. Kalak, J. Beck, T. Götz, T. Schmidt, et al. 2012. "Facing Depression with Botulinum Toxin: A Randomized Controlled Trial." *Journal of Psychiatric Research* 46:574–581.

Woodward, J. 2003. *Making Things Happen: A Causal Theory of Explanation.* Oxford: Oxford University Press.

Woodward, J. 2010. "Causation in Biology: Stability, Specificity, and the Choice of Levels of Explanation." *Biology & Philosophy* 25 (3): 287–318.

Woodward, J. 2015. "Interventionism and Causal Exclusion." *Philosophy and Phenomenological Research* 91 (2), 303–347.

Ylikoski, P. 2013. "Causal and Constitutive Explanation Compared." *Erkenntnis* 78 (S2): 277–297.

Zangwill, N. 2012. "Constitution and Causation." *Metaphysica* 13 (1): 1–6.

Zednik, C. 2011. "The Nature of Dynamical Explanation." *Philosophy of Science* 78 (2): 238–263.

Zhang, M., and X. Li. 2012. "From Physical Weight to Psychological Significance: The Contribution of Semantic Activations." *Journal of Consumer Research* 38 (6): 1063–1075.

Zhong, C. B., and G. J. Leonardelli. 2008. "Cold and Lonely: Does Social Exclusion Literally Feel Cold?" *Psychological Science* 19 (9): 838–842.

Zhong, C. B., and K. Liljenquist. 2006. "Washing Away Your Sins: Threatened Morality and Physical Cleansing." *Science* 313 (5792): 1451–1452.

Zhong, C. B., B. Strejcek, and N. Sivanathan. 2010. "A Clean Self Can Render Harsh Moral Judgment." *Journal of Experimental Social Psychology* 46:859–862.

Ziemke, T. 2003. "What's That Thing Called Embodiment?" In *Proceedings of the 25th Annual Meeting of the Cognitive Science Society*, edited by R. Alterman and D. Kirsh, 1305–1310. Mahwah, NJ: Lawrence Erlbaum.

Zlochower, A.J., and J. F. Cohn. 1996. "Vocal Timing in Face-to-Face Interaction in Clinically Depressed and Nondepressed Mothers and Their 4-Month-Old infants." *Infant Behavior & Development* 19:371–374.

Index